Praise for Theatre of the Mind

"Ingram works his magic of translating scientific research into the coinage of common understanding. He deftly outlines dozens upon dozens of experiments that lift the curtain to reveal a glimpse of consciousness, the last remnant of the soul. . . . Best of all, Ingram, as always, delivers the goods in a manner that is readily accessible—and a lot of fun."

—*The Globe and Mail*

"An amazing, clever, witty book. Jay Ingram is a brilliant author and scholar. *Theatre of the Mind* is simply a must-read."

—Alan Kingstone, Ph.D., Departments of Psychology and Neuroscience, University of British Columbia

"Lively . . . witty . . . in the tradition of Stephen Jay Gould and Oliver Sacks. [Ingram] manages a difficult trick—making the minutiae of science seem alluring to the uninitiated."

—*Maclean's*

"Jay Ingram takes us on a guided trip down the stream of consciousness. . . . An enjoyable jaunt."

—Bernard J. Baars, Ph.D., Saybrook Graduate School and Research Center

"Ingram's language is user-friendly and easy to understand, the ideas seductive and the science thought-provoking. . . . Should you read this book? A qualified yes is the answer. . . . If you are interested in how your mind experiences reality, this is a layman-friendly review of how little we know and how error-prone our perceptions are."

—*Winnipeg Free Press*

"[*Theatre of the Mind*] abandons the episodic essay format for an ambitious, in-depth examination of a single topic: human consciousness. . . . Ingram recounts thought-provoking experiments, anecdotes and research . . . that suggest how far we've come along in our understanding of consciousness and how far we've yet to go."

—*Ottawa Citizen*

JAY INGRAM

Theatre of the Mind

Raising the Curtain on Consciousness

HARPER
PERENNIAL

Published by Harper Perennial, an imprint of HarperCollins Publishers Ltd

Originally published in hardcover by HarperCollins Publishers Ltd: 2005
This trade paperback edition: 2006

HarperCollins books may be purchased for educational, business, or sales promotional use through our Special Markets Department.

HarperCollins Publishers Ltd
2 Bloor Street East, 20th Floor
Toronto, Ontario, Canada
M4W 1A8

www.harpercollins.ca

Library and Archives Canada Cataloguing in Publication

Ingram, Jay
Theatre of the mind : raising the curtain on consciousness /
Jay Ingram—1st pbk. ed.

ISBN-13: 978-0-00-639455-6
ISBN-10: 0-00-639455-8

1. Consciousness. I. Title.

BF311.I54 2006   153   C2006-903973-9

RRD 9 8 7 6 5 4 3 2 1

Printed and bound in the United States

The author gratefully acknowledges permission to reprint the following images: 38 courtesy James T. Rutka, M.D.; 41 courtesy Moonrunner Design Ltd.; 54 reproduced from John C. Eccles, "Figure 8.9," *Evolution of the Brain: Creation of the Self:* 190, with permission from Thomson Publishing Services on behalf of Taylor & Francis Books; 85 courtesy Lydia Kibiuk; 89 courtesy Daniel J. Simons; 92 courtesy Dr. Ron Rensink; 93 courtesy of Al Fenn/Getty Images; 150 courtesy Moonrunner Design Ltd.; 164 courtesy Prof. Brock Fenton; 179 reprinted from Nicholas Humphrey, "Cave Art, Autism and the Human Mind," *Journal of Consciousness Studies* 6 (1999): 40–41, with permission from Elsevier; 209 reprinted from Jeffrey W. Cooney and Michael S. Gazzaniga, "Neurological disorders and the structure of human consciousness," *Trends in Cognitive Sciences* 7, 4 (April 2003): 162, with permission from Elsevier; 242 reprinted from Marcus E. Raichle, "The Neural Correlates of Consciousness: An analysis of cognitive skill learning," *Philosophical Transactions: Biological Sciences* 353, 1377 (1998): 1894, with permission from the Royal Society. All other images created by HarperCollins Canada.

To my family

Contents

Introduction

I MAGINE what it would be like to live in a completely different world, an alien place in which you couldn't even know your own mind, a place where, bombarded by sensory stimuli, your mind could extract only the merest hint of the total, and yet you somehow believed — or were made to believe — that you were all-seeing and all-knowing.

Imagine further that this was a world where you were, in effect, an automaton, a self-deluded one at that, that walked and talked and acted at the behest of a whole set of mental modules whose operations you knew nothing about. Imagine that one of those modules, entirely of its own volition, routinely took the sparse data available to it and concocted bizarre stories to explain what was happening.

Think of a world where an array of photons bombarding the retina becomes a three-dimensional space full of objects and textures, which can then be transformed to an internal image of the same scene, which later, when the brain is in a completely different chemical state, can be brought to mind once again, but this time in a wholly foreign setting.

The strangest thing about this world is that seemingly magical transformations take place: ideas, visions, hallucinations and memories are created by electricity and chemistry, crossing the boundary between the physical and the immaterial in a manner that defies logic and science. The mystery is deepened by the fact that tinkering with the chemistry, the electrical circuitry or even the physical substrate of the brain can dramatically change those thoughts and images, without shedding any light at all on how they are created.

Well, look no farther. This is exactly the world that you now inhabit. It is the world of the conscious mind, a world that is not at all what it seems. The science of consciousness has, in the last two decades, transformed our thinking about the brain and how it creates the world you experience.

There are actually several Theatres of the Mind. One is the little theatre in your head that seems to be the place where the events in your conscious mind run, like a never-ending film. It is the home of your mind's eye. It is a theatre with an audience of one: you. Only you can tell the rest of us what's playing.

It's one thing for us to feel that way, but sometimes that inner theatre concept worms its way into people's minds so thoroughly that even researchers seem to be envisioning it when they discuss consciousness. Philosopher Daniel Dennett has spent years skewering this idea of what he dismisses as the "Cartesian theatre." He argues that consciousness is not being processed, edited and presented to anyone/anything in the brain. There is no anyone/anything—consciousness is the end of the line.

He's been effective. Consciousness experts take pains to dissociate themselves from any notion of an inner theatre, but at the same time, one of the most popular theories of how consciousness works, called the "global workspace," is best viewed as—what else?—a theatre, with consciousness being played out on the stage. But it is being played for no one: it just is.

Illuminating the theatre of the mind has become, in the last few years, one of the most challenging and exciting areas of science. That is what this book is about.

CHAPTER ONE

What Is Consciousness?

THERE is something unnerving about writing a book about a subject that even the experts can't define. Welcome to consciousness. It really shouldn't *be* that hard: we may not think much about consciousness, but we all know what's it like not to have it. When you are sleeping—as long as you're not dreaming—you are *un*conscious. If you faint or get knocked out, you *lose* consciousness. In those situations, as far as you know, there is nothing much going on in your head. Actually, there is much going on in your brain, but you are unaware of it. Awareness and consciousness are closely related, if not synonymous.

Getting knocked out or slipping into deep dreamless sleep are the most common excursions from consciousness, but there are others. If you have a general anaesthetic for surgery, you are unconscious as long as the anaesthetic is in effect. At least, most people seem to be, but there are unnerving reports from patients like this one: "I could see the surgeons at the end of the operating table and I thought, 'O my God, they are going to operate on me and I'm awake.' I tried to tell them but I couldn't speak—couldn't move . . . it was the worst experience of my life." Another wrote, "The consciousness was terrifying. . . . The desperate animal terror of trying to signal one's conscious state to someone, but unable to twitch a bloody eyelash."[1]

Less dramatically, patients can sometimes remember or recognize certain words that were spoken while they were "under" the anaesthetic. In extreme cases, derogatory comments made about the patient ("This one's today's Blue Plate special") or the alarmed reaction of the surgeons ("Gee, this could well be a tumour . . .") have been linked to delayed postoperative recovery, suggesting that the words were heard in some sense, though they apparently never actually reached consciousness. Sometimes post-operative hypnosis does reveal that the patient did hear and can remember comments made in the OR.

But does this suggest that there is a kind of consciousness while under anaesthesia? Not necessarily. Because muscle relaxants that induce paralysis are given alongside the anaesthetic, patients suffer, as in those real-life accounts, but also are unable to signal in real time that they remain aware, if indeed they are. As a result, scientists are forced to rely either on indirect indicators of consciousness (such as patterns of brain waves) while the patients are apparently knocked out or on the patients' memory afterwards, jogged by hypnosis or not. Neither provides the kind of hard and fast reliable data you'd like. This is the subjectivity problem that arises again and again in consciousness research: my thoughts and feelings are my own. No matter how cleverly and vividly I'm able to describe them to you, that description is necessarily incomplete, my consciousness at least partly inaccessible to you. For all we know we are all completely aware during surgery but forget it all after. Who can prove that wrong?*

What about those momentary lapses during freeway driving, in which you suddenly realize that you have no memory of the last

---

* While on the subject of the unprovable, let's raise—and then dismiss—solipsism, the idea that it's all complete subjectivity. Solipsists (if indeed there are any) claim there is nothing except your mind, and that it has created everything else you are aware of. Woody Allen said it best: "What if everything is an illusion and nothing exists? Then I definitely overpaid for my carpet." One of the funniest unintentional remarks about solipsism was contributed by Bertrand Russell, who reported that a woman had written him saying she was a solipsist and that she was surprised by that fact that there were no others (Bertrand Russell, *Human Knowledge: Its Scope and Limits* [London: Allen and Unwin, 1948]).

ten kilometres? They seem at first glance to be a short-lived version of the same sort of thing, but while you may not remember your driving, you weren't unconscious. You were talking on your cellphone, listening to the radio or lost in other thoughts, and each of those requires that you be conscious and attentive to them. The fact that your mind had "wandered" from the routine of driving to other, more interesting things is a striking example of the ability of consciousness to select or reject whatever mental activities it wishes. It also illustrates how a thought or image can move easily into consciousness or back out of it, taking only fractions of seconds to do so.* That's what happens to your awareness of driving, at least for a while.

This doesn't necessarily mean you were accident-prone during that time on the road, although how would you ever be able to find out? Any sudden emergency would snap you back to full consciousness in an instant, and even if it didn't, it likely wouldn't make any difference: your foot would already be on the brake, directed there unconsciously (more about this later). Even if your momentary lack of awareness did cause you to react marginally more slowly, you'd likely never know. What would you have to compare that moment with? It's also possible that once you noticed something was happening and snapped back to full awareness mode, any memory that your mind had been wandering might have been erased.

Bearing in mind that the unconscious is close at hand at all times, it might be easier to define consciousness as its flip side, the entire set of things that goes missing while you are unconscious. But what are those things exactly? An unending series of images and words—even music—running through your "mind"? The ability to re-create some remote or no-longer-existent room in a house of years past, complete with colours, aroma and

---

* It's also true that while you can talk to your passenger and drive unconsciously, you can't do the reverse—talk unconsciously and pay attention to the driving—suggesting that language and consciousness are more closely linked than driving and consciousness.

sound? The capacity to think about the fact that you are thinking? The answer is all of the above and more.

A common moment, rich with conscious impressions.

Let's build it up slowly: you're sitting at the breakfast table, it's
early morning, the sun is streaming in the window, coffee's in a
cup, there's food and a newspaper—even this simple, quiet scene
provides a bewildering load of sensory information to cope with.
How well does your stream of consciousness (likely psychologist
and consciousness pioneer William James's most famous phrase)
accommodate this ever-changing flood of sensory data plus the
appropriate memories, beliefs and emotions and incorporate
them all into consciousness? Pretty well for most of us—we feel
as if we have an absolutely complete and high-fidelity awareness
of the breakfast-table surroundings.

And beyond that? The sky's the limit. Imagine you're adding
milk or cream to your coffee or tea. How do you judge how much to
add? Usually by the colour. How do you do that? By comparing
your memory of the colour of previous cups to the changing colour
of this one, ready to stop pouring when they match. It definitely

requires consciousness—you can't add the cream to your coffee while experiencing one of those freeway blackouts. You must pay attention to, and be aware of, the coffee.

Easy to say, but where does that memory of the previous cup's colour come from? How do you bring it back from wherever it normally rests into your conscious mind, so that you can stop pouring at just the right time? Where is that normal resting place? One simple memory like that inevitably prompts other, deeper ones, such as the great cup of coffee you once had in Banff on a cold fall day. These are not just dry thoughts—they bring with them the emotions of the memories as well as those of the moment.

So far the breakfast-table scene has omitted one important ongoing activity of consciousness. While every sip of coffee or turn of the newspaper page reveals fresh new information to be incorporated into consciousness, there is, at the same time, an ongoing awareness of yourself. You know that it is you at the table, and not someone else. You are aware that while thousands of people might be enjoying the morning sun, only you are enjoying it *your way*. It's not actually an additional responsibility for your consciousness; maintaining the identity, privacy and subjectivity of it all is actually inseparable from the thinking, feeling and noticing part. If nothing's running through your mind, there's no self there to notice.

But consciousness is not just about content, rich though it is. There is also a quality to it, feelings that go with it, something that *it is like* to be conscious. If you thought consciousness was hard to define simply, the feelings of consciousness are worse. Philosophers call them *qualia*, and such feelings are, as far as anybody knows, absent from other nonconscious information-processing entities, such as computers, or even small-brained animals or insects. But they are definitely part of consciousness. What are they? Here I can only defer to those who think about them. The philosopher Frank Jackson writes, "Tell me everything physical there is to tell about what is going on in a living brain, the kind of states, their functional role, their relation to what goes on at other times and in other brains, and so on and so forth, and be I as clever as can be in fitting it all together, you won't have told me about the hurtfulness of pains, the itchiness of itches, pangs of jealousy, or about the characteristic experience of tasting a lemon, smelling a rose, hearing a loud noise or seeing the sky."[2] David Chalmers, one philosopher who has famously labelled the final explanation of these experiences the "Hard Problem," adds a similar list: "the felt quality of redness, the experience of dark and light . . . the sound of a clarinet, the smell of mothballs . . . the experience of a stream of conscious thought."[3] Finally, Christof Koch, a neuroscientist, comes up with a similar, but more involved pair of examples: "the raw experience of the haunting sad vibes of Miles Davis's 'Kind of Blue' or the ecstatic, near-delirious feeling of dancing all night."[4]

I quoted these examples to show you that the experts know what qualia are even if I'm not so sure. My problem is this: if my awareness of the colour red is already packaged with the feeling of "redness," then I can't imagine what it would be like without that feeling. Anyway, you can sort this out yourself and even add your own examples: what it's like to eat marshmallows roasted over a campfire, or to feel the heat of that campfire on your face. Many philosophers, including the two above, Jackson and Chalmers, think qualia will be the showstopper for the scientific effort to understand consciousness, that these feelings will simply be impossible to explain on the basis strictly of brain activity. And the puzzle of qualia goes beyond the "how" to the "why." Why does your brain create the sensation of a really bad burger, when it should be enough to enumerate the long list of chemicals, the feel of soggy bun and chewy meat, the texture-without-taste of the tomato slice, then add those sensations up and throw it away? What good does it do us to have the burger "experience"?

There is another side to this. There are many dissident philosophers and scientists who aren't daunted by qualia. Some argue that there really isn't anything extra there, and that by the time consciousness itself is understood, qualia will automatically be revealed as part of the package. Others agree that the experience, the feelings, of being conscious are a mysterious sort of add-on, but they are content to wait until those feelings are easier to explain, that is, until consciousness is much better understood. In their view, qualia, even though unexplainable now, may prove to be understandable with a few more years' worth of knowledge.

What might qualia be for? Psychologist and vision expert Richard Gregory argues that qualia are crucial to brains like ours that are loaded with memories and anticipated futures as well as with the complexity of the now.[5] Of those, only the now comes equipped with the feelings associated with qualia, so everything that is immediate and relevant is more vivid. Gregory (and others) argue that qualia "flag the present" as the situation to be dealt with now. In the absence of such a flag, the present, the past and the future would be hopelessly confused. Simpler animals have no such need of qualia because their lives are mostly stimulus-response, with little thinking in between.

You can see how this might be true. The memory of the taste, appearance or texture of an apple is a pale imitation of the sensation you get when you're actually looking at one or biting it. The memory lacks the vividness (the qualia) of the real experience and as a result is easy to recognize for what it is. The real apple, the one we have to deal with in the here and now, stands out from the rest. It has been highlighted. Gregory has suggested the following experiment: look carefully at something distinctive, such as a bright red tie, then shut your eyes and witness the disappearance of its vividness. Then reverse the experiment, starting with your eyes closed, imagining the tie, then opening your eyes: as Gregory notes, "the perception is strikingly vivid compared with the imagination or memory."

It's not just memory: I think that any process or medium that separates you from an experience, however high-fidelity that medium is, degrades the qualia of that experience. Television is a perfect example. You can have a seventy-two-inch plasma screen, high-definition television with theatre-quality sound, it doesn't matter. Imagine watching a documentary on the exhilaration of sailing. Is it comparable in any way to being on deck, feeling the spray on your face, the tilt of the boat, the wind? No. I had a very different but just as telling experience when I visited MD Robotics in Toronto, makers of various paraphernalia for the space program, including the Canadarm and Canadarm 2, incredibly complex robotic manipulators. I had already seen hours of footage of the arms in action, both real and animated, but that didn't prepare me for the sensation of actually standing next to one. They are more beautifully engineered, complex, shiny, massive, unbelievably impressive when you see them in the flesh—their qualia shine through in a way that only being in their presence can provide.

So the fact that our memories and some images are not as colourful as the experiences that gave rise to them helps the brain attend to things that are of immediate importance, but beyond that, if our memories were as vivid, as laden with qualia, as the real world, they would be indistinguishable from it, and that would make them hallucinations.

As I say, I still find them hard to understand. The sound of a clarinet, to me, is part of its identity—the defining part, in fact. It's a challenge to dissect that from the experience of what it's like to hear a clarinet. In case you're like me, here are a couple of last-ditch stabs at making clear exactly what qualia are.

First, neuropsychologist V.S. Ramachandran with a cross-species example: "Imagine there's a species of electric fish in the Amazon that is very intelligent, in fact as intelligent and sophisticated as us. But it has something we lack: the ability to sense electric fields, using special organs in its skin. You can study the neurophysiology of this fish and figure out how the electric organs on the sides of its body transduce electric current, how this is conveyed to the brain, what part of the brain analyzes this information, how it uses this information to dodge predators, find prey, and so on. If the electric fish could talk, however, it would say, 'Fine, but you'll never know what it *feels* like to sense electricity.'"[6]

That might work for you, but if not, here's another analogy: the fruit industry could develop a robot able to identify just when peaches were perfectly ripe and ready to eat. It could enumerate every single chemical constituent of the peach, compare those to the ideal numbers of a perfect peach and then decide just how close to perfection was the peach it was analyzing. But I think you can appreciate that that robot would not experience a fresh peach as you or I would. (On the other hand, some experts in artificial intelligence would argue that if that robot were as complex, processor-wise, as the human brain—laid out in a comparable way—then that robot could well experience the peach.) Maybe a raccoon would enjoy that peach, while an ant wouldn't. Whoever or whatever has it, that experience appears to be an additional quality of consciousness.

That is what consciousness is, roughly speaking. But it wouldn't be reasonable to stop there, because there are other forms of consciousness, some of which we suspect, others that we have actually experienced. One of those that we all have enjoyed—or not—is dreaming. This is a crazy form of consciousness, full of the images that we experience in waking life and more, but linked together in a totally implausible way. There is no sensory input during dreams—the brain just works with its memory, its capacity for creating images and its emotion, and the result is totally convincing, no matter how absurd it seems later to the awake brain. Waking consciousness resorts to a certain amount of storytelling to make the series of events in our lives plausible, but dreaming takes this craft to a whole new level (or abandons it completely!).

Altered consciousnesses can result from the mental discipline of meditation or from mind-expanding drugs. Both can shatter some of the most basic aspects of normal consciousness, such as the feeling that you are you, bounded by your skin and separate from your surroundings. All these different versions of consciousness are important because they offer new perspectives for both the person experiencing them and the scientists who study them. But unless you're pretty unusual, you spend most of your time in the daily grind of waking consciousness, and it is that state that I'll investigate from the scientist's point of view.

It's really only in the last ten or fifteen years that scientists, especially neuroscientists, have taken consciousness seriously as a scientific subject, something that can be experimented with, something that, while subjective at its core, still provides opportunities for gathering objective scientific data. The vast majority of these scientists believe that the mind and the brain are one, that once we understand enough about what exactly the brain is doing when we're conscious, we will know what consciousness is. And even though some of them suspect we might never get the final answer, they're willing to work at it until stopped dead by some ultimate roadblock.

Of course, it's possible that the brain and the mind are two different things entirely (a position defined as "dualism"), the mind being something unique, immaterial not physical, yet connecting somehow to the physical brain such that coma and sleep interrupt it. There are very few neuroscientists who admit to being dualists. Most believe that consciousness will be explained by studies of the brain in action, but even those who aren't sure — who might be open minded about the possibility that the mind represents some new kind of matter or energy — are tackling those parts of the problem that they think can be dealt with.

The Conscious Thermostat

Okay, this is not a joke. The idea that something as simple as a thermostat might be conscious has been discussed among philosophers. One of the most prominent philosophers of consciousness, David Chalmers, has proposed that information is the basis of consciousness, and by that standard, even a thermostat (which is, after all, capable of change in response to variation in information) is conscious. I should add that Chalmers doesn't really think the thermostat in his house "feels" the differences in temperature.

Chalmers was scooped on the idea of the idea of the conscious thermo-stat by John McCarthy, of the computer science department at Stanford University.[7] McCarthy argued that it should be appropriate in some cir-cumstances to assume that a machine of some kind had beliefs, as a sort of shorthand way of understanding what it might do, or what state it was in, where an absolute and complete understanding of its inner workings might not be easily figured out. Obviously, the more complex the machine, the more appropriate assigning beliefs to it would be. Even so, McCarthy has something to say about thermostats:

First consider a simple thermostat that turns off the heat when the tempera-ture is a degree above the temperature set on the thermostat, turns on the heat when the temperature is a degree below the desired temperature, and leaves the heat as is when the temperature is in the two degree range around the desired temperature. . . . We ascribe to it the goal, "The room should be ok." When the thermostat believes the room is too cold or too hot, it sends a message saying so to the furnace. . . . We do not ascribe to it any other beliefs; it has no opinion even about whether the heat is on or off or about the weather or about who won the battle of Waterloo. Moreover, it has no introspective beliefs; i.e. it doesn't believe that it believes the room is too hot.

And if that doesn't at least open the door to your believing that ther-mostats could have beliefs, try this:

The temperature control system in my house may be described as follows: Thermostats upstairs and downstairs tell the central system to turn on or shut off hot water flow to these areas. A central water-temperature thermostat tells the furnace to turn on or off thus keeping the central hot water reservoir at the right temperature. Recently it was too hot upstairs, and the question arose as to whether the upstairs thermostat mistakenly *believed* it was too cold upstairs or whether the furnace thermostat mistakenly believed the water was too cold. It turned out that neither mistake was made; the down-stairs controller *tried* to turn off the flow of water but *couldn't*, because the valve was stuck. The plumber came once and found the trouble, and came again when a replacement valve was ordered. Since the services of plumbers are increasingly expensive, and microcomputers are increasingly cheap, one is led to design a temperature control system that would *know* a lot more about the thermal state of the house and its own state of health.

There are even bizarre imaginary characters lurking in the consciousness landscape. The possibility that consciousness is something beyond science opens the door for philosophers to invoke an imaginary being called a "zombie" to argue their position that mind and the brain are completely different. Picture beings who are identical to you, right down to the molecules, their atoms and even the parts of their atoms, who behave to all appearances exactly as you would, but who have no inner life, no awareness, no consciousness. They are robots, automatons. Such zombies would provide living proof of the difference between mind and brain, because he would have all the brain you have, but without the mind. It would be difficult to hold to the faith that brains like ours inevitably produce consciousness if theirs didn't. Putting zombies like this into play hasn't deterred those who don't buy into dualism, at least partly because they are able to counter that if indeed this creature had exactly the same kind of brains we do, it would be conscious— inevitably. Simple as that.

While there aren't very many posters of zombies on the walls of neuroscience labs, there is one individual who is scorned even more, and that is the "homunculus." This is an imaginary little guy (there is never any attempt to feminize him) who sits in an easy chair in your brain watching a big-screen HDTV that displays all your sensory information, processing it all and thinking your thoughts. It's too bad he's so unpopular, because he'd be the perfect explanation for the fact that consciousness seems to hang out at one central point behind your eyes. That would be the place for the "theatre of the mind," the screen, the easy chair, the Ikea rug with the taco-chip crumbs on it. (In fact this notion is usually derided as the "Cartesian theatre," a slap at René Descartes for being bold enough to suggest there was a specific place in the brain—he chose the pineal gland—where mind and brain meet.) Unfortunately, if you need a little man in your brain to explain consciousness, then he needs one in turn, and so on and so on, creating an infinite set of nested *matryoshka* homunculi.* That never-ending, explaining-nothing quality of the homunculus has made him *persona non grata* in consciousness science. The alternative, as we'll see in more detail later, is that consciousness is distributed across the brain—that there is no one place where it happens. That is a beautiful idea because in one fell swoop it deals with both zombies and homunculi, by replacing the latter with a series of fairly stupid, definitely unconscious (but industrious) "zombie agents." We will meet them later.

It isn't easy to study consciousness scientifically, partly because of the complexity of the brain, partly because of the subjectivity thing. I know I'm conscious. How do I know *you* are? (For that matter, how do you know *I* am?) We both look and act like we are, but because consciousness is subjective, because there's no way I can peek into your brain and check (no amount of brain imaging can make what you're feeling visible to me), all I have is your word that you are, and vice versa. Not only that—even if we both are conscious, I can never know whether your consciousness and mine are the same. (It's the old puzzle: you might call a certain colour orange, and I'll agree, but we can't know if your orange and mine seem the same to each of us. Maybe your orange is my green and vice versa.)

It's not just that your experiences are literally beyond description, but even the testimony you provide as to how you're feeling or what you're thinking is of questionable value. Now, that statement should bring you up short. Of course we know what we are thinking! How could we not? But there is, unfortunately, abundant

---

* Ironically, the original homunculus of centuries ago was put together the same way. He was the miniature individual thought to exist inside very human sperm—a "pre-formed human"—who upon fertilization became the embryo. Naturally, the sperm homunculus had to contain its own miniature testicles, full of miniature sperm containing even tinier pre-formed humans, with the series ending with the last human that would ever be born. This was clearly a theory that was hard to confine to science.

evidence that we really have no good insight into our innermost thoughts.

Two experiments should convince you. The first, called "Reasoning in Humans" was conducted by psychologist Norman Maier in 1931.[8] Maier placed volunteers in a large room containing a variety of objects, such as poles, clamps, pliers, extension cords, tables and chairs. He then hung two cords from the ceiling, each of which just reached the floor, just far enough apart that no one could grasp the end of one and at the same time be able to reach the end of the other. The challenge was to figure out how to use the objects in the room to tie the ends of the two ropes together.

Most of the test subjects stumbled on the easiest solutions almost immediately. These included tying one cord to a chair, moving the chair as far as it would go towards the other cord, then grabbing the second cord and tying them together; lengthening one cord by tying the extension cord to it or hooking one cord with the pole and reeling it in. But there was one solution that most of the participants overlooked, so Maier gave them a hint: he walked casually by one of the cords and set it swinging. Within a matter of seconds—in almost every case in less than a minute—sixteen subjects saw the light. They tied a weight onto the end of one cord, set it swinging and then ran to the other cord and held it until the swinging cord approached close enough to grab. The timing of the idea was important: even though it took most of them less than a minute after the hint, they had been stumped by the problem for anywhere from five to ten minutes leading up to it.

And how many of those sixteen subjects who took the hint realized that that was exactly what had happened? One. One! The other fifteen made up stories about their solution of the problem. They claimed they didn't notice the swaying cord, they hadn't realized there was a hint, hadn't even seen the experimenter.

So then how did they think it had happened? "It just dawned on me"; "Perhaps a course in physics suggested it to me." One subject, a professor of psychology, caricatured his entire profession by saying, "Having exhausted everything else the next thing was to swing it. I thought of the situation of swinging across a river. I had imagery of monkeys swinging from trees. The imagery appeared simultaneously with the solution. The idea appeared complete."

If that experiment weren't enough, here's another. This, one of my favourite examples of the inability to recognize what goes on in our heads, was a study performed by social psychologists Richard Nisbett and Timothy Wilson.[9] Nisbett and Wilson set card tables up in a shopping mall with a sign that read, "Institute for Social Research—Consumer Evaluation Study—Which Is the Best Quality?" They then set out four identical pairs of panty hose (Agilon, cinnamon) and asked passers-by which pair was the best. No one had a problem choosing the best of four identical pairs, and no one had a problem identifying what it was about the pair he or she chose that made it the best. Among the reasons: knit, weave, sheerness, elasticity and workmanship. Fifty people proffered a total of eighty reasons, but none identified the real reason: the position effect. Reading from left to right, the pairs, labelled A through D, were preferred by 12 per cent, 17 per cent, 31 per cent and 40 per cent of the participants. When it was suggested that placing might have been an influence, all but one denied that possibility. That one individual, who was taking three psychology courses at the time, identified the crucial importance of positioning . . . but chose B. The point is that she *chose*, when there really was no choice to make. These two experiments demonstrate beyond a tiny shadow of a doubt that we cannot trust our own analysis of our own thinking. And if *we* can't, who can?

So even though we can't really be sure, we operate on the assumption that you are conscious, that I'm conscious and that we likely have roughly the same kind of consciousness. But is human consciousness truly all the same kind?

There's no evidence that males and females are consciously different, despite the differences in men's and women's brains. Of course when a man and a woman are exposed to the same events, they'll have a different conscious experience—a different experience, but not a different consciousness, in the way that I'd imagine a Neandertal person's consciousness might be different from ours, or a chimp's different from both. But I'm not sure I'm on solid ground even there: the trend in experiments with primates of many kinds is that they're more sophisticated than we thought, and many of the old divisions between Neandertals and modern humans (the people formerly known as Cro-Magnon) have broken down: the Neandertals' brains were bigger than ours today, they made sophisticated tools (not just crummy handaxes) and they survived a daunting variety of climate changes. While fifty years ago it was believed that we either killed them off or beat them in straight-ahead competition, now the best guess is that there was one final climate change that the Neandertals couldn't cope with, and it extinguished them. Did Neandertals have a different consciousness than we do? I'd bet they did, although it's hard to imagine what it would have been like.

Of course, further back in time our ancestors were less like us: much smaller brained, likely inarticulate and less inventive. They would likely have had a consciousness, but not a *Homo sapiens* version. How far along that line can we go? Most scientists today would credit the great apes with being conscious, and some would include cats, dogs and even most mammals, but lines would begin to be drawn at frogs, snakes and insects. With rare exceptions, no one would grant consciousness to bacteria.*

That is a rough outline of what consciousness is, or appears to be. But the really important thing about it is that your consciousness is yours and yours alone. The philosophers and neuroscientists may be studying it, but only you know what it's like. Before we take the plunge into the science of consciousness, we need to

---

* It depends on who you talk to, or at least who you might have talked to a century ago. Even well into the twentieth century there were a few brave biologists who felt that protozoa, such as amoebae and paramecia, were actually conscious, that their movements reflected active thought and consideration rather than mindless reactions to chemicals in their environment. These biologists were by no means a majority, but they represented a current of thought that had lasted for a hundred years.

take the time to reflect on exactly what it seems to be . . . to you. The chances are that you haven't really given it much thought, but the next chapter encourages you to do that. Then, and only then, can we look at the scientific approach to consciousness, and the place where it's assumed to live—in the brain.

CHAPTER TWO

The Feel of Consciousness

F OR more than ten years now, a group of Swiss researchers led by Dietrich Lehmann has been recording the electrical activity in the brain in the hopes of identifying patterns, visible signs of the thoughts going on in that brain.[1] This is a difficult task, given that any electrical signal that can be detected represents the joint activity of millions of brain cells, their precise location muddied by the fact that the signal has had to pass out of the brain, through the skull, to the recording electrodes. But even so, Lehmann has found some intriguing things. For instance, he and his team have identified what they call "microstates," bursts of neuronal activity that last only about 100 or 200 milliseconds, bursts that apparently can be traced to different circuits of neurons. More important, these microstates appear to correlate with the experiences the person is having.

In the most intriguing experiment, student volunteers were asked to sit and let their minds run free. They were told only that a tone would sound periodically, and as soon as they heard it they were to report what they had just been thinking. They were given no other instructions (the less going on in the brain the better). Lehmann and his group found that it was possible to classify brain activity as either visual or abstract. For instance, an imagined beach scene qualified as visual, while a student who apparently was mulling over the meanings of the word "belief" just before the tone sounded provided an example of abstract thinking. (He also qualified as an unusual undergrad.)

Once those thoughts were classified, they were then compared to the two full seconds of EEG activity immediately previous to the tone. The amazing thing was that abstract thoughts correlated with a brief shift of activity to the left side of the brain, but only in the last couple of hundred milliseconds before the tone—the EEG record from two full seconds before the tone revealed no such bias in activity. In other words, the thoughts reported by these subjects had apparently been in existence only for a tenth of a second or a little bit more. In fact, when the researchers pooled their results, they concluded that, at least in this experiment, a thought lasted about 120 milliseconds.

The researchers concluded that "the seemingly continuous stream of consciousness consists of separable building blocks which follow each other rapidly" and then wondered why, if this is indeed the case, we nonetheless think of consciousness as a steadily flowing set of thoughts, images and feelings. Maybe that's just the way we experience mental events. They likened it to a movie, where two hours of cuts, dissolves and flashbacks seem to us, if well done, to have unfolded seamlessly. It could be the same for the thoughts generated in our brains.

Is there such a disconnect between what we experience as our own consciousness and what it really is? There's not much doubt that consciousness *seems* continuous. While you are undoubtedly aware at any time of most of the objects and events surrounding you and of the thoughts and feelings inside you, they don't stand still; they arrive, pause, then leave in sequence, a never-stopping flow. William James, whose writings a century ago still serve as the cornerstone for ideas about consciousness, coined a phrase to describe this feeling that thoughts are always on the move: the "stream of consciousness." James likened it to a "river, forever flowing through a man's conscious waking hours." It's not just a term for psychologists: writers like Saul Bellow and James Joyce put their streams of consciousness on paper, and Bob Dylan sometimes put his to music. It's also a term that has been borrowed by the Internet: "streaming video."

But the stream of consciousness is a good place to start taking a closer look at what consciousness feels like. The problem is that most of us don't think about it at all, and while trying to define the way it feels may not be much of a scientific approach to the subject, it at least establishes what the scientists are chasing. After all, that *is* their target: the inner workings of the mind. *Your* mind.

First, is consciousness really a stream, a steady flow of mental images, thoughts and feelings? Even William James qualified his own imagery by arguing that the stream was in some ways like the flight of a bird:

"Like a bird's life, it seems to be an alternation of flights and perchings. The rhythm of imaginations of some sort, whose peculiarity is that they can be held before the mind for an indefinite time, and contemplated without changing; the places of flight are filled with thoughts of relations, static or dynamic, that for the most part obtain between the matters contemplated in the periods of comparative rest."[2]

To me that makes consciousness sound like some sort of Rube Goldberg machine, in which a ball falls onto a trampoline, which sends it back up into a chute, from where it rolls down an incline to strike a pendulum, which swings back and forth, and so on. There are changes of pace, short and long pauses—the forward movement of a stream with none of the steadiness. The virtue of this analogy is that it preserves the craziness that sometimes best characterizes consciousness. Both James's birds in flight and a Rube Goldberg machine tally with the results of Dietrich Lehmann's experiments. If microstates are indeed "thoughts," then the shift in electrical patterns from one to the other represents the fluttering of consciousness's wings or the rolling of its ball.

There are data that more directly represent the contents of the progress of consciousness. For decades psychologists have been trying to explore daydreaming, surely the purest form there is of consciousness in motion. However, it's tricky to pin down: it's easy enough to simply allow someone to sit still and think, but the challenge is to get inside their minds.

One method is to ask participants to press a button every time their thoughts change (analogous to Lehmann's tracking of their electrical activity). But how do you know that their thinking isn't changed by the very fact that they are obliged to track their thoughts and act on them? It's the old dilemma originally posed by the ideas of quantum mechanics in physics: as soon as you observe the phenomenon, you've changed it. "Thought sampling" is another technique, in which a bell or some other signal prompts the person to relate what they were thinking at that moment. This, of course, relies on the person's memory, and some investigators worry that the detail of consciousness, what has been called the "sequential fine-grain," is lost. A third approach is thinking out loud. It too has its faults: people doing this seem to linger longer on each thought than they might have if they hadn't been giving voice to them. They expand on content where thought sampling shrinks it. However, thinking out loud does have the virtue of revealing the stream, the bouncing ball, whatever you want to call it.

Here are some examples from actual transcripts of thinking out loud from a couple of decades ago at Yale University:[3]

It's hard to draw, hard to draw many parallels . . . tile floor, our floor in the bathroom where I live . . . which looks like a movie, I don't know why, *Fantastic Voyage* occurs to me. Looks more like Mexico. Socks, shoes, blue jeans, pillow, thread, white and blue thread, white walls, tires, 1950, California, Allen Ginsberg, beards, hair, black hair, red hair, Lucille Ball, television, Bob Hope, nose, lustful, Nathaniel West, H. L. Smith, half-dollars, six dollars and fifty cents, uh George Washington, Abraham Lincoln, and Abraham Lincoln's birthday, Valentine's Day, candy, cookies, sugar, white sugar, red sugar, brown sugar, white opium, white heroin, brown heroin, *Time* magazine, *Newsweek* magazine, *Life* magazine, death magazine, in-between magazine, uh, television, uhm, radio, red socks, blue ones, ball games, popcorn, hot dogs, mustard, yellow, uh, John Lennon, uh, glasses, lots of glasses, Elton John . . .

Now that's a stream! Here's another example, utterly different and yet so similar:

I mean ultimately what I would love to do is be in New York, trying in one way or another in live theater to make my living through it, but [name of person deleted] and I keep planning how when we get to Hawaii just don't do anything that's going to lead to anything. Maybe I'll go central casting for *Hawaii Five-O*. No, but I'd rather be in California. Any time there's a day like today, I can't imagine why anybody would choose to stay on the East Coast anywhere from October to February or March. If Yale wasn't here, I mean, I'm glad I'm at Yale, I wouldn't want to be anywhere else only because of, the only reason I would want to go anywhere is the climate. If this could be moved to California that would be ideal.

My socks are too short. I ate too much at lunch. . . . Where is my mind? I feel like it's at a standstill, only because I'm supposed to be saying something. This isn't very helpful. . . . I wish I could . . . I think so much during the day when I'm thinking that I'm not supposed to be thinking that I can't concentrate on what I'm supposed to be concentrating on. Lucky I didn't go to class today. Yet if I sat there I would have looked out the window thinking about the work I'm supposed to be doing, thinking about how I'm not going to get it done and how I'd like to go dancing this weekend if there was a dance. And yet now when I'm sitting here and don't have to do anything but just think cause I love to think, I'm not thinking. . . . I should have my hair cut. I'm getting bored with it.

These examples are unlikely to represent a typical minute or so in anyone's life, partly because they were produced under experimental conditions (as is obvious from the second monologue) and partly because most of us have little time to let our brains run free. But they are a window into what's going on.

In fact, there is good experimental evidence that our brains are at least as active as those transcripts would suggest all the time. A French research group, rather than asking volunteers to participate in specific mental tasks, asked them to think of *nothing*. It's hard to imagine being able to do that, but the participants were supposed to keep their eyes closed, remain immobile and avoid engaging in mental tasks like counting. When the subjects' brains were imaged, they turned out to be more active under these conditions than in any one of nine demanding tasks, such as the perception of words or mental calculation. The subjects themselves reported that their mental rest period was absolutely *filled* with thoughts. The brain images matched this testimony, in that it revealed activity of brain areas thought to be involved with memory, imagery and emotion.

So consider the transcripts as examples of what our brains can do, given the chance, but also as pieces of a more complex mosaic of consciousness that is typical of daily life.* A little reflection (bearing in mind how inaccurate *it* can be) suggests what other pieces of that mosaic might be like.

What about the song that keeps running through my head whenever I'm not concentrating on writing this (at the moment it's Dylan's "One More Cup of Coffee")? It has been there in my brain for about two hours now, surfacing only when . . . well, I'm not sure what prompts or permits its intrusion, but it seems as if it lurks there, just offstage, ready to reveal itself at a moment's notice. We've all had the experience of discovering that the song that happens to be running through our minds—often a song that we haven't thought about for months, if not years—appears to have some eerie connection to something that has just happened in our lives. I'm sure that sometimes it is just coincidence,

---

* In fact, it's been argued that daydreaming and real dreaming have much more in common than we might think—that daydreaming only seems paler, less dramatic, more mundane, because it has to compete with an incoming stream of sensory information. Dreaming is immune to that input. However, there's no direct evidence that daydreaming *is* dreaming while awake—it just resembles it.

sometimes the very common association of love song with love life, but sometimes it seems to be a case of a song surfacing from the unconscious, apparently unbidden.

That idea that there is some kind of pre-consciousness, a place where ideas wait in the wings, explains another common experience. If you're juggling a number of ideas and things to do, and one by one you deal with them, you can still be sure when you've momentarily forgotten one because it leaves you with, say, an uneasy feeling. Even though you're not directly *aware* of it—you can't name the event or situation or person that has made you feel that way—you still know that it's there. It is a strange turn of events: you are sure there is something there in your consciousness, but hidden, like the stuffed animal under the blanket, betrayed only by the lump it creates. Funnily too, when you finally do identify it, the feeling that it created subsides, replaced by the awareness of what it is. The feeling might still be there, not completely replaced but overshadowed by the precise and direct spotlight of attention you can direct at it. I'm not aware of any research that would substantiate the existence of some staging area for consciousness, but that is what it feels like. This might be what William James called the "fringe" of consciousness: "the mere feeling of harmony or discord, or a right or wrong direction in the thought."

There are other examples of this partial or fringe consciousness. One is the tip-of-the-tongue phenomenon. Most of the time when you can't think of a word, you can nonetheless know something about it. The first syllable, or at least the first letter, the length of the word and maybe even the last syllable are familiar (it's called the "bathtub" effect because it reminds you of a very tall person in a bathtub, with his head and feet visible as they hang over the ends but the all-important middle hidden in the tub). James alluded to the same sort of thing when he wrote,

Suppose we try to recall a forgotten name. The state of our consciousness is peculiar. There is a gap therein; but no mere gap. It is a gap that is intensely active. A sort of wraith of the name is in it, beckoning us in a given direction, making us at moments tingle with the sense of our closeness, and then letting us sink back without the longed-for term. If wrong names are proposed to us, this singularly definite gap acts immediately so as to negate them. They do not fit its mould. And the gap of one word does not feel like the gap of another, all empty of content as both might seem necessarily to be when described as gaps.[4]

Imaging of the brain when it is stumped by the tip-of-the-tongue phenomenon has shown that parts of the brain involved in the resolution of conflict—that monitor the recruitment of suggested solutions and evaluate them for correctness—are, not surprisingly, active when subjects are struggling to find the word. Two further points about fringe consciousness: First, the next time you have a tip-of-the-tongue experience or any other fringe consciousness moment, examine what it feels like, and I think you'll find that the fringe lacks qualia—it doesn't have that intensity of experience that full consciousness has. Second, it's sometimes true that the unconscious can come to your rescue in situations like this. If you are in the middle of reciting a prepared statement of some kind (or, even better, singing a song) and you fear that you have a memory blank coming up, sometimes it's just better, rather than thinking about it, to plow ahead, and often your unconscious will find the word for you.

And while few of us have the opportunity to record the meanderings of our stream of consciousness in the same way the students at Yale did, most of us *do* carry on a running commentary in our brains that, in the minds of some experts, is an important part of consciousness. Alain Morin of Mount Royal College in Calgary is one of those who argues that something called "inner speech" is what allows us not just to be conscious, but to think about it.[5] Inner speech is the unvoiced talk that runs through our brains (in some cases, at least from what I can tell from asking people, virtually continuously).* Sometimes it breaks through the silence when we talk to ourselves, but much of the time it's just there, sometimes competing with a tiresome song, but always providing a running commentary on mental events. Of course, reading this activates your inner speech—you can't read without it.

Thinking about your own consciousness is self-awareness. Morin contends that it's one thing to be conscious—to be seeing, smelling, thinking, feeling and remembering—but that it's another to think about those elements of consciousness, to say to yourself, "That's the smell of my grandmother's kitchen when she was baking," or "I'm so bored—I wonder when I'm going to be able to get out of here." Inner speech, according to Morin, amplifies the stream of consciousness, making it more vivid, but is not necessary for consciousness itself. There are cases of brain damage where inner speech has disappeared but the patient is clearly still conscious, although perhaps in a diminished way. Inner speech is what allows you to build an image of yourself; any time you have to fill out a questionnaire that asks you to describe your personality traits, you have to rely on inner speech, because it is what you use to describe what goes on in your mind.

I pointed out in Chapter 1 that the famous example of loss of consciousness—"Where did the last ten kilometres of highway go?"—is mislabelled. It's not that you weren't conscious, it is that you were conscious of other things than driving. I've come to believe that there is a good reason for this. Have you ever reflected on what's it like to pay attention when you're driving? It's exceedingly boring! Try it next time you're on a highway: you move your eyes from the inside rear-view mirror to the outside one, check out the tail-lights of the cars ahead of you, glance at the gas gauge and speedometer, move your eyes from one rear-view mirror to the next, check out the tail-lights of the cars ahead of you . . . it's a case of letting your mind wander or going insane. It's also true that rou-

---

* Philosopher Bernard Baars claims that you can't stop your inner speech for any longer than ten seconds, and that he's never been able to do it himself for more than five.

tines like driving require less and less conscious attention as they are learned better and better—"overlearned," as the psychologists say. So once you're behind the wheel, the act of driving is relegated to neural circuits that don't require consciousness, and unless your mind is going to go blank, you begin to think of other things.*

Consciousness also seems different to me in the first few moments of the day. Waking sometimes begins with a blank slate that, rather than being filled in instantaneously, builds up piece by piece: I know I'm awake, I know where I am, I can glance at the clock and know what time it is. But at this point the list of things that I have to do today still hasn't fallen into place. These features, which for the rest of the day will be, if not top of mind at every instant, at least hovering nearby, are added the way the icons at the bottom of my computer screen appear on booting up: one by one. However, after these few brief moments consciousness seems complete, the screen is full, and it continues that way until the next night of sleep.

Each of the above examples touches on one aspect of what it's like to be conscious. There of course could be many more, some of which would be common to us all, some of which might be something you feel that I don't. As a final thought I want to raise an idea that might seem strange in the extreme, but only because it is a question that is seldom asked: why is it that my mind seems to be centred just behind my eyes, right in the middle of my head?** As we will see, consciousness is a highly processed and abstracted version of the world outside the head, an invention more than an impression; if so, why must it feel as if it is between my ears? Why couldn't it be floating somewhere above my head, or behind it, or just off to one side?

---

* Is it even possible for your mind to "go blank"? In some versions of Zen there is a state called "no-thoughtness." It is said that if you're able to achieve this state, you can be a spectator as a thought arises, bursts into consciousness, then fades and eventually disappears, giving rise to no other thoughts.

**The physicist Erwin Schrodinger, in a little but prized book called *What Is Life?*, argued that consciousness feels as if it's "an inch or two behind the midpoint of the eyes" (*What Is Life? The Physical Aspect of the Living Cell* [Cambridge: Cambridge University Press, 1945]).

But why, you say, should it? After all, it happens in the brain, and that's where the brain is! More accurately, it's between my eyes and ears, and why not have your "self" in close proximity to our two predominant senses, vision and hearing, senses that possess the ability to localize the sights and sounds that they are perceiving? If an object is "over there," it makes sense to relate over there to the eyes and ears, and so to the centre of the head, and to do it as efficiently as possible, especially if that object is dangerous.

However, your awareness of sounds and sights likely isn't generated in the brain areas where those sensory impressions first arrive — most of these probably never reach your consciousness at all, but those that do have moved around as they were processed, sometimes between places that are far enough apart that you would have to point to two different places on your skull to locate them. That being the case, why shouldn't consciousness at least shift around a little from moment to moment?

For whatever reason, or maybe there's no reason at all, consciousness portrays itself as anchored in the middle of your head. I should admit that while this is true for most of us, surveys have revealed that there are other locations: Susan Blackmore, in her book *Beyond the Body*, recorded that while most people place their self behind the eyes, "some say the middle of the forehead, the back of the head, or even the throat or heart."[6] She goes on to point out, as many others have, that at least when it comes to memories, our selves are often displaced. If you think back to the last time you were on a beach, for instance, chances are you see yourself from behind, from above or from just about anywhere except from inside your head.

Even if it turns out there is a central brain location for consciousness (and there's a lot of skepticism that anything like that will ever be found), there would be no obvious reason for it to go to the trouble of announcing its location. After all, it spends much of its time and imagination travelling in space and time — why centre those images in the middle of the head? The late Julian Jaynes, for my money one of the most inspirational commentators on consciousness, wondered the same thing: "To think of our consciousness as inside our heads . . . is a very natural but arbitrary thing to do. I certainly do not mean to say that consciousness is separate from the brain; by the assumptions of natural science it is not. But we use our brains in riding bicycles, and yet no one considers that the location of bicycle riding is inside our heads. The phenomenal location of consciousness is arbitrary."[7]

But it does make sense to have it in your brain and not your foot doesn't it? Today it does. But in Aristotle's time, for instance, the heart was thought to be the centre of the intellect. Did people then imagine their selves to be somewhere just inside the ribcage? Did they think that their eyes and ears were periscopic in nature, sampling the world from an elevation? We'll likely never know, because the opinions of the ancient Greek on the street weren't preserved by history. (Of course, we can speculate, as Jaynes did in his book *The Origin of Consciousness in the Breakdown of the Bicameral Mind*, which will appear again in Chapters 5 and 11. Jaynes wondered if because decision making can be stressful enough to cause breathlessness and a pounding heart, would it have been natural to suspect that those decisions were being made in that part of the body?)

Are there people today who experience their self as being somewhere else in the body?* Patients who have had significant brain surgery furnish some clues here: among these patients are those with the corpus callosum, the cable of fibres transmitting information between the two cerebral hemispheres, severed to confine the wildfire spread of their epileptic seizures to one hemisphere, rather than both. Their daily lives are to a large extent unaffected by the operation, but in the lab, where images and sounds can be directed to one hemisphere or the other, their deficit can be dramatically demonstrated (see Chapter 13). Even

---

* After talking about this on television once, I received an e-mail from a man with dissociative identity disorder (formerly called multiple personality disorder) who feels as if "much of the time I spend outside of my skull, a few inches in front of my face. When I am going to sleep it can shift (almost as though it is retracting) back into my skull."

though each of the two hemispheres seems to possess its own style of thinking, language capability and even knowledge, the patient's consciousness is unaltered. Each gives testimony to the same sort of inner self as anyone with intact channels between the hemispheres. Even the speech hemisphere, which apparently speaks only for itself and not its twin, never argues that it is seated behind the left eye rather than being in the middle (the speech hemisphere is almost always the left). Yet studies of split-brain patients have shown that each of the two hemispheres has its own consciousness, its own set of ideas, its own actions based on those ideas. That being the case, why doesn't our consciousness at least switch from one side to the other depending on what hemisphere is most active at any moment?

Even more dramatic are those rare cases in which a young child requires the removal of an entire hemisphere, as a last-ditch treatment for severe epilepsy or encroaching brain disease. In this case it is not a question of which hemisphere might argue that it is the site of consciousness—there *is* only one. Yet there are no reports of anyone having had a hemispherectomy who reports that their self is squeezed to one side or the other in their skull, although from what I've been told, the question has never been asked.

If it's true that the apparent location of consciousness has more to do with the sense organs than with the brain itself, maybe that location would change if those senses were somehow compromised. Helen Keller is a dramatic example: deprived of both sight and hearing, her principal sense was touch, and the main organs of touch her hands. There is no doubt she was conscious, but there is no reference that I'm aware of as to where she experienced that consciousness.

Out-of-body experiences—OBEs—appear to be beautiful examples of cases where the self has escaped its cranial prison, but they don't have to be. In the standard OBE (if there is indeed such a thing), the person feels as if her mind has left her body and may be simply floating above it or, in extreme examples, travelling somewhere else. Obviously, the travelling aspect is paranormal and way beyond the issue of whether consciousness can be experienced somewhere other than in the middle of the skull, but the more straightforward cases of feeling as if you're looking down at yourself—with nothing else changed—seem to be good examples of exactly that. The question is, how reliable are these reports?

Charles Tart of the Institute for the Scientific Study of Consciousness has studied out-of-body experiences for years, and has reported some amazing cases.[8] In one, a subject named Miss Z. apparently had OBEs quite often while asleep and was actually able, in one such experience, to read a randomly generated number that Tart had written on a slip of paper high on a shelf beyond her reach:

Each laboratory night, after the subject was lying in bed, the physiological recordings were running satisfactorily, and she was ready to go to sleep, I went into my office down the hall, opened a table of random numbers at random, threw a coin onto the table as a means of random entry into the page, and copied off the first five digits immediately above where the coin landed. These were copied with a black marking pen, in figures approximately two inches high, onto a small piece of paper. Thus they were quite discrete visually. This five-digit random number constituted the parapsychological target for the evening. I then slipped it into an opaque folder, entered the subject's room, and slipped the piece of paper onto the shelf without at any time exposing it to the subject. This now provided a target which would be clearly visible to anyone whose eyes were located approximately six and a half feet off the floor or higher, but was otherwise not visible to the subject. . . . On the fourth night, at 5:57 a.m., there was a seven minute period of somewhat ambiguous EEG activity, sometimes looking like stage 1, sometimes like brief wakings. Then Miss Z awakened and called out over the intercom that the target number was 25132, which I wrote on the EEG recording.

The number 25132 was indeed the correct target number. I had learned something about designing experiments since my first OBE experiment and precise evaluation was possible here. The odds against guessing a 5-digit number by chance alone are 100,000 to 1, so this is a remarkable event! . . . Unless Miss Z, unknown to us, had employed concealed apparatus to illuminate and/or inspect the target number, which we had no reason to suspect, there was no normal way for anyone lying in bed, and having only very limited movement due to the attached electrodes, to see it.

Of course Tart, being interested in demonstrations of the paranormal, wondered whether cases like that of Miss Z. really meant that the mind could leave the body, or whether these experiences were simply hallucinations. (They appeared, in at least Z.'s case, not to be dreams.) Either way, the mind doesn't necessarily have to occupy the head, except for the possible reason I suggested earlier, that it's good from the point of view of survival to have your mind right there behind your eyes and between your ears. That is exactly the point Tart makes:

"To jump to the end point of my and others' researches, it is useful to see our ordinary consciousness as a process that creates an ongoing, dynamic simulation of reality, a world model, an inner theater of the mind, *a biopsychological virtual reality* 'in' which consciousness dwells. . . . We should first realize that the ordinary feeling that we are 'in' our bodies (usually our heads) is a construction, a world *simulation*, that happens to be the optimal way to ensure survival most of the time, but that it is not necessarily true in any ultimate sense."

Tart goes on to suggest that:

". . . it is helpful to remember that, just as a person using a high quality computer-generated virtual reality simulator forgets where their physical body actually is and becomes experientially located 'in' the computer-generated world, it might be that our 'souls' are actually located on Mars, but we are so immersed in the BPVR (biopsychological virtual reality) our brains generate that we think we are here in our bodies. This is a crazy idea, but helps to remind us that the experience of where we are is not a simple matter of just perceiving reality as it is."[9]

Few would disagree with that last sentence—or at least with its first five words. A more down-to-earth view of OBEs has come from cases in which patients with electrodes implanted in their brains (for the treatment of epilepsy) experienced their own OBEs when certain areas of the brain were electrically stimulated. The analysis of such cases has focused on the importance of those brain areas for the perception of our body parts and their positions. But one thing's for sure: if you can gaze down at your body from a point above it, something has happened to the apparent location of your mind. A "crazy idea" for sure, but one that raises questions about consciousness—not just why we experience it where we do, but also what exactly it is and, more relevant to most of us, how different consciousness is from the way it feels. To begin to answer that question, at least from the scientific point of view, we have to know something about the place where it all starts: the brain.

CHAPTER THREE

Is Your Brain Really Necessary?

I used to think my brain was the most fascinating part of the human body, but then I thought, "Look what's telling me that."

—Comedian Emo Phillips

T HE title of this chapter is lifted from a provocative television program made in 1980. It featured the late Dr. John Lorber, a pedia-
trician in Manchester, England, and a small group of his patients.
Lorber's specialty was hydrocephalus—water on the brain—a condition in which an infant suffers from an excess of cerebrospinal
fluid in the skull, as a result either of overproduction of the fluid or
of inadequate drainage. If the fluid cannot be removed and the pressure is unrelieved, normal development is seriously compromised.
The still-malleable skull bulges outward, but, more critically, the
brain inside is under incredible pressure and can't grow properly.
Before surgically implanted shunts were introduced in the 1960s to
relieve the fluid pressure, only about 20 per cent of children with
hydrocephalus reached adulthood, and half of those were brain-
damaged. Dr. Lorber was a pioneer in the implantation of small
plastic shunts to rid the brain of fluid and give such infants a new
lease on life. However, that legitimate claim to fame was overshadowed by something much more controversial. Lorber stated that

some of his patients, despite having suffered what should have been crippling brain damage, had apparently shrugged it off and were leading normal, or even better-than-normal lives. Here is what he said in that program:

> There's a young student at this university who has an IQ of 126, has gained a first-class honours degree in mathematics, and is socially completely normal. And yet the boy has virtually no brain. . . . When we did a brain scan on him we saw that instead of the normal 4.5 centimetre thickness of brain tissue between the ventricles and the cortical surface, there was just a thin layer of mantle measuring a millimetre or so. His cranium is mainly filled with cerebrospinal fluid. . . . I can't say whether the mathematics student has a brain weighing 50 grams or 150 grams, but it is clear that it is nowhere near the normal 1.5 kilograms, and much of the brain he does have is in the more primitive deep structures that are relatively spared in hydrocephalus.

In the television program *Is Your Brain Really Necessary?* Lorber reiterates that the student's brain amounts to "not more than 5 per cent" of the weight of a normal brain. We are shown CAT scans of the Sheffield math student's skull, and there is a vast empty space occupying what seems to be most of the upper half. Hardly any brain at all: a millimetre of brain tissue clinging to the inside wall of his cranium. "Virtually no brain . . . no detectable brain" is how Lorber puts it. He then displays a model of the man's brain, revealing the great hole in the middle. This is the miracle: a man of above-average intelligence who has achieved that with almost no brain.

Yet very few serious brain scientists took Lorber seriously at the time, or were even familiar with the story. That is understandable: one of Lorber's accounts appeared in the British magazine *The Nursing Mirror* in 1981, another in a 1983 German book called *Hydrocephalus im frühen Kindesalter*, in which he claimed to have examined 150 cases of extreme loss of brain tissue, 30 or so of whom had not been prevented from attaining above-average IQs. There was also the film, but nothing in any journal where the manuscripts are reviewed by experts. There was a brief account in the news section of *Science* magazine, by science writer Roger Lewin, but nothing official. Nevertheless, Lorber's work is still widely quoted—especially on the Internet—as evidence that the mind and the brain are two different things and that mind can be intact even when the brain is definitely not.

Lorber was a little prickly in his retirement years. I know, because I talked to him once on the phone. I was working on a documentary series on the brain (*Cranial Pursuits*) for CBC Radio[1] in the early 1990s, and when I reached him at home and asked if we could talk about his research, the first thing out of his mouth was a demand for £150 sterling before he'd even speak to me.

Lorber was defensive, and I can see why. Some prominent neuroscientists to whom I mentioned Lorber's work just dismissed it out of hand: as far as they were concerned it just had to be wrong. If Lorber was right, most of the last century's brain research would just go down the toilet. For that reason alone you would expect that they would dismiss it out of hand—they'd have to. The paradigm is that the brain creates the mind, and as everyone who has actually read Thomas Kuhn's *The Structure of Scientific Revolutions* knows, the ruling paradigm doesn't admit contrary evidence gracefully. The paradigm doesn't usually change until all its defenders are dead. However, there is another good reason that neuroscientists have ignored Lorber. His evidence is weak, largely unsubstantiated and, more important, *all the rest of the evidence* says that you *do* need your brain.

In the television documentary, we are shown two images corresponding to horizontal slices through the brain of one of Lorber's three extraordinary patients and are told that the two slices are at "slightly different levels," but we are not told how far apart those "slices" were. If the difference between them was only a millimetre or two, then there's no way of knowing how far the empty space extended in either direction vertically. This is obviously a crucial issue that would have been resolved if there were some serious, peer-reviewed treatment of the data—something beyond a television documentary. It is true that these images reveal that the ventricles, the fluid-filled hollows present in every brain, have expanded dramatically because the fluid contained within them was pushing outward. Indeed, other images are presented of the math student showing the same thing. But in none of these pictures do we see a layer of brain tissue as thin as a millimetre crushed up against the inside of the skull.

Lorber then reveals a model of one of these hydrocephalic brains, this time sliced vertically, not horizontally. There is the predictable empty space, but again we have no idea how far the space extends forward or back. It is impossible to calculate from these snapshots just how much brain tissue is missing.

A third part of the documentary takes Dr. Lorber and his patients to Copenhagen for a different kind of brain imaging, one that identifies what parts of the brain are active. These images suggested that some remarkable reorganization had gone on, including shifting the visual areas of the brain to one side from their usual location at the back. What isn't mentioned in the documentary, but is evident from the scans, is that the amount of brain tissue present and active—perhaps 50 per cent—is far more than Dr. Lorber is claiming.

It's odd that the critics apparently haven't taken the time to scrutinize the documentary, or they might have found these inconsistencies. Instead they have mostly resorted to rude dismissals or the pre-emptory "Oh, imaging was so much cruder and misleading then." Recently one of Lorber's former colleagues supported that contention by saying, "If the cortical mantle actually had been compressed to a couple of millimeters, it wouldn't even have shown up on his x-rays."[2] Another claimed that magnetic resonance imaging has shown that hydrocephalic brains are pushed and pulled out of shape during development but don't actually suffer much loss of tissue.

If that latter statement is true, the Lorber claims are rendered even less spectacular. It might sound like I'm splitting hairs here, but the actual amount of brain tissue remaining is a crucial issue when claims are being made that we don't need our brains or that we have vast reserves of unused brain tissue that can come into play in case of devastating accidents or illness. Specifically, the loss of, say, 40 to 50 per cent, while sounding unbelievable, is in the same ballpark as the loss sustained in a hemispherectomy, the removal of one entire side of the brain.

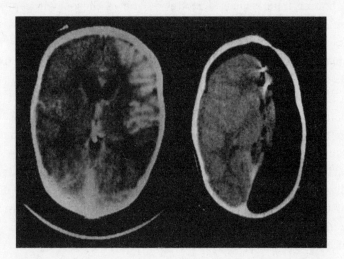

Patients who have had hemispherectomies function well despite having lost a considerable amount of brain tissue. As the image on the right shows, they are missing an amount of brain about equal to that lost, according to John Lorber, by some children who had had hydrocephaly. Yet Lorber claimed that his patients should force us to question whether we really need our brains to be able to think.

Hemispherectomies vary in the actual amount of brain tissue removed, but there are several cases where an entire hemisphere was removed—not just the upper half, the cerebral hemispheres, but even the structures paired along the midline below. Obviously, this is a surgery of last resort, when a brain tumour can be stopped only by removing the entire hemisphere that it is occupying, or when epileptic seizures are just too powerful and disruptive.

Hemispherectomies often leave the patient disabled in some way (although not as disabled as they would have been if the operation had not been performed). Sometimes, though, these operations not only halt the course of disease, but at the same time allow full and normal mental development.

The point is that there are people who have defied the odds and enjoyed a full mental life even after suffering the loss of a significant amount of brain tissue. Whether their success is the result of a dramatic reorganization and reallocation of neural resources or even of new growth of the brain is not clear. But either mechanism, or a combination, explains the phenomenon without having to resort to the claim that there are vast areas of the brain that aren't used except in such an emergency, or that our minds are actually independent of our brains. The answer to Lorber's question, "Is Your Brain Really Necessary?" is "Yes."

While we're on the subject of how much brain we need, let's beat up on that old myth that "We use only 10 per cent of our brains."

It's been around for nearly a century. The percentage varies— I've heard as much as 25 per cent—but regardless of the number, the implication is that there are vast areas of our brains just sitting there idle, waiting for us to call on them. There is no evidence for this today, and there never has been any, so where did the idea come from and why is it still around?

Barry Beyerstein, a psychologist at Simon Fraser University, has painstakingly searched for the origin of the idea and has likely come as close to finding it as anyone will. In a chapter he wrote for a book called *Mind Myths*, Beyerstein traced the first clear reference to the 10 per cent claim to a lecture delivered by William James to the American Philosophical Association in December 1906. But what James actually said was, "We are making use of only a small part of our possible mental and physical resources," and the rest of the lecture made it clear that he was talking not about a largely silent brain, but about the myriad factors—including fatigue and even social convention—that prevent us from thinking freely and energetically all the time.

It's possible that there is no unambiguous origin for the 10 per cent idea (I found the same to be true for the claim that scientists have "proven" that the bumblebee cannot fly), but there is no doubt that in the 1920s and '30s it became widely quoted. The most influential apostle of the idea was Dale Carnegie, author of *How to Win Friends and Influence People*. In 1944 he wrote, "The renowned William James was speaking of men who never found themselves when he declared that the average man develops only ten percent of his latent mental abilities."[3]

Now, James's original statement and even Carnegie's reference to it can be read several ways. If they are arguing that most of us, plagued by the dull necessities of daily life, feel there are greater things we could accomplish, there'd be no argument. But the modern interpretation of this idea is clearly that there's just a lot of brain doing absolutely nothing. Why has this mutated version of the idea established itself? Part of it is that we all like to think we could probably lead a more active mental life. It's also true that believers in strange and unproven phenomena, especially of the mental kind, can point to the unused 90 per cent and claim that is where it all could happen, if only you could train yourself to use it.

But unfortunately there is no evidence whatsoever that there are unused parts of the brain. On the contrary, brain damage, sometimes involving only tiny amounts of brain tissue, almost invariably leaves some deficit in its wake. If the 10 per cent argument were true, you'd hear doctors say, "He's lucky he suffered that stroke in the idle 90 per cent of his brain." But you never do.

Brain imaging techniques have revealed that if enough mental and physical tasks are performed, every single cubic centimetre of brain tissue will be recruited, much of it more than once. Obviously, if you're not riding a bicycle right now, those circuits in your cerebellum that have trained up for that aren't as active as they would be if you were, although I'd bet that if you *imagined* riding a bike, they'd respond.

I could go on: When the great Canadian brain surgeon Wilder Penfield let his electrode wander over the exposed surface of his patient's brains, he found no vast unresponsive areas; evolution would hardly have encouraged the growth of an organ that requires 20 per cent of the body's oxygen if 90 per cent of that organ had no function.

Lorber's patients and the 10 per cent idea are likely the two most popular claims to trot out when someone is trying to argue that the brain isn't necessary for the mind or consciousness. There are philosophical arguments too, as we'll see, but brain science today is based firmly on the idea that the brain is what it's all about. And if you're looking for consciousness in the brain, and nowhere else, you'd better know the territory.

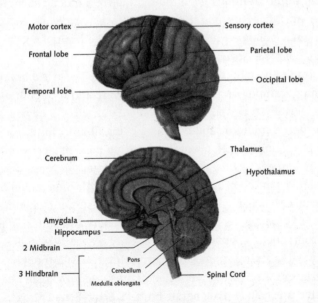

The cerebral cortex, the outer covering, dominates any view of the human brain. The cortex is so highly folded that it is impossible to see, without dismantling it, just how voluminous it is. Above, the left hemisphere intact. Below, the inner surface of the right hemisphere exposed by removing the left hemisphere.

The human brain weighs about a kilogram and a half, is a little rubbery, has deep wrinkly folds over most of its surface and is laced with ropy blood vessels, and the view from above reveals that it is clearly divided into a right and a left half. From this vantage point you're seeing the famous right and left hemispheres of the brain, the halves that experience such different lives—lives of which you are totally unaware.*

Even though the two hemispheres are mostly separate already, if you were to split the brain all the way down the middle and examine the newly exposed inner surfaces, you would see that you had not only severed a major connecting cable between the two hemispheres, the corpus callosum, but had also split in two a number of structures nearer the floor of the brain. These are small, somewhat independent organs, at least when it comes to function: responsible for different things, even though they're all made of the same brain stuff. They include the thalamus, a major connection point for signals going in various directions in the brain, and the hippocampus, the structure that got its name because someone apparently thought it looked like a sea horse. There is a branch of the hippocampus on each side of the brain, and if both branches are surgically removed or irreparably damaged, the owner is left with the crippling disability of not being able to form any new memories; time passes, events happen, but none of it is recorded.

Why, you ask, would anyone perform surgery to remove such a crucial part of the brain? It happened in 1953, when a man code-named H.M. became the most famous brain surgery patient of all time. H.M. was in his late twenties at the time and suffering terribly from repetitive epileptic seizures. Surgery was used then—as it is now—to remove injured or scarred brain tissue thought to be the trigger for seizures. Renowned neurosurgeon William Beecher Scoville was called in to operate on H.M. Scoville had already performed several of these hippocampectomies on people with schiz-

---

* See Chapter 13, "Split Brains."

ophrenia or bipolar disorder, and the operations had had the desired effect of calming them. In H.M.'s case the surgery was a success . . . sort of. His seizures ceased, but ominously, from the day after his surgery, H.M. was clearly having trouble remembering things. He couldn't find his way back to his hospital room; he didn't recognize people he had met only hours or even minutes before. It soon became clear that H.M. had lost all ability to make new memories. He was stuck in 1953.

And he is still stuck today. He can remember details of his childhood in Hartford, Connecticut, but he draws a blank with anything that has happened since 1953. He has no idea how old he is, where he is, who the president is or what year it is. He is still mentally agile, able to complete crossword puzzles (as long as the clues require no post-1953 knowledge), but he is, in many ways, a lost man. If told once again of the death of his favourite uncle (which happened decades ago), he is suddenly saddened, but five minutes later he has forgotten once more. Eerily, he has the notion that important things can be learned from him because of some surgery that went wrong.*

The cruel irony is that had there been the right kind of follow-up of Scoville's schizophrenic patients, he would never have conducted such radical surgery on H.M. The problem was that the goal of the surgery was to pacify the schizophrenics and make them more manageable, which it did. Nobody bothered at the time to check their memories. When that was done after the H.M. debacle, it was clear that they too had become stuck in the past. With that knowledge, Scoville would likely have opted for a less dramatic procedure.

Brains that have been damaged inadvertently have also played a hugely important role in revealing how the brain is organized. One remarkable example was discovered in the early 1980s in Germany. A woman was admitted to hospital in a stupor. She had

---

* H.M. has other fragments of memory that relate to events after his surgery: he knows, for instance, that a U.S. president was assassinated, but he can only guess at who he was or where it happened.

been suffering from headaches, nausea and vomiting, and an examination revealed that she had suffered a stroke that had damaged the back of her brain, the area that processes visual information. What was unusual was that her vision had been disrupted in only one place—a small area at the back of the brain called MT—and in only one way: she could no longer perceive movement. This is what the doctors reported:

> She had difficulty, for example, in pouring tea or coffee into a cup because the fluid appeared to be frozen, like a glacier. In addition, she could not stop pouring at the right time since she was unable to perceive the movement in the cup (or a pot) when the fluid rose. Furthermore the patient complained of difficulties in following a dialogue because she could not see the movements of the face and, especially, the mouth of the speaker. In a room where more than two other people were walking she felt very insecure and unwell, and usually left the room immediately, because "people were suddenly here or there but I have not seen them moving. . . ." She could not cross the street because of her inability to judge the speed of a car, but she could identify the car itself without difficulty. "When I'm looking at the car first, it seems far away. But then, when I want to cross the road, suddenly the car is very near." She gradually learned to "estimate" the distance of moving vehicles by means of the sound becoming louder.[4]

A follow-up exam years later showed that her disability persisted, although she had learned some coping mechanisms.

That the hippocampus and the vision region called MT have specific roles to play is typical of the brain. It is full of regions, structures and circuits that on their own do one specific thing. There is a face-recognition area, a place-recognition area, places for the perception of speech and for the production of speech, areas that guide your hand when you reach for the coffee cup, a different area that enables you to name the coffee cup (or call it up in your mind's eye), areas that map touch sensations to the appropriate place on your body, areas that control your movements, and many, many more. The brain is "modular." There's no agreement on exactly how modular it is, because scientists are a long way from mapping the entire brain and knowing what each part does. But so far it looks like the whole brain is the sum of its parts.

This sounds a little bit like nineteenth-century phrenology, where the brain was carved up into areas responsible for character traits such as cautiousness, wonder, wit, tune and vanity. There was one additional feature in phrenology: if one of those parts of the brain was unusually well developed, it was supposed to bulge, creating a bump in the skull immediately above. So a person's character could be read by feeling those bumps. Phrenology is seen today as an absurd sort of neurological mass delusion, but maybe we shouldn't be quite so quick to ridicule it—after all, attributing precise mental functions to specific locations in the brain, which was phrenological in the nineteenth century, is modular in the twenty-first.

Let's move now from this rough overview of the brain to the other end of the scale, the ultimate building blocks of the brain: the brain cells, the neurons. Neurons are signalling devices that rely on electricity and chemistry to do their job. They are spidery cells, with a central part, the cell body, less than a tenth of a millimetre in length, but with extensions that can reach several millimetres or even centimetres. Typically there is one extraordinarily long process called an axon, and a myriad of shorter ones called dendrites, which branch and re-branch into a tangled mesh. The axon carries the nerve impulse to other cells; the dendrites accept impulses from other cells. In the right orientation a neuron looks like an ancient tree in dire need of pruning.

The impulse itself is an electrical wave that sweeps down the length of the axon. When it reaches the end, the impulse can go no farther, because there is a gap (an infinitesimally small gap) between it and the adjacent neuron, a gap across which it cannot pass. However, if the impulse is strong enough it causes hundreds of tiny sacs containing thousands of molecules called neurotransmitters to migrate to the inner surface of the cell, fuse with it, then discharge their contents into the gap. If enough of the transmitters make it across the gap (and it's only one fifty-thousandth of a millimetre across) and plug into the mirror-image receptors that stud the surface of the neuron on the other side, an impulse can be created in that cell. The electrical wave travels down the axon of the neuron at around 5 to 10 metres per second, which, while on the scale of the brain is pretty good, isn't blindingly fast.

The space between neurons, called the synapse, is a prime target for drugs of all descriptions. One of the best examples are the Prozac-like drugs, the SSRIs, selective serotonin reuptake inhibitors. Here's how they work: in some mental disabilities, like depression, there are abnormally low levels of the neurotransmitter serotonin. A nerve impulse might release packets of it, but not enough to stimulate the next cell in the sequence. Normally the serotonin that is released is quickly mopped up by an enzyme, removed from the synapse and transported back into the neuron that released it. This is ordinarily a sensible thing to do: if neurotransmitters were free to float around in the synapse, they would subject the receptors to a never-ending bombardment and that cell would become hyperactive.* So reuptake, as it's called, is crucial to normal brain function but not helpful when there's too little serotonin to begin with. Prozac-style drugs interfere with the reuptake mechanism, forcing the serotonin molecules to stay in the synapse and raising the probability that

---

* Speaking of hyperactive, most psychedelic drugs have molecular structures similar to serotonin and likely find their way to the synapse and bind to those same serotonin receptors. Obviously, combining a drug like LSD with an SSRI would be a fairly stupid thing to do. Also, leeches employ serotonin to regulate their feeding; the more of the neurotransmitter, the more aggressive they are. This has led me to fantasize about a cottage owner spilling his summer's worth of Prozac into the water beside the dock, then jumping in for what would turn out to be a truly memorable swim.

they'll find a receptor and actually stimulate a nerve impulse. It is a roundabout way of making sure that a nerve impulse that should be carried forward actually will be.

This is a very simple picture: there are hundreds of different neurotransmitters. Some have three or four different receptors they can plug into; sometimes a single neuron is capable of emitting more than one kind; some are excitatory—they stimulate the neuron next in line—but some are inhibitory, and there are, without doubt, subtleties to the way they work that haven't even been discovered yet. But these are the basic units from which the brain is built. I introduce them this way because from here on the picture becomes much, much more complex.

Just take the numbers. There are likely 100 billion neurons in the brain, about as many neurons as there are trees in the Amazon rainforest. Impressive enough, but each of those neurons forms as many as *ten thousand* synapses with other neurons. Suddenly what seemed to be a relatively simple and not particularly fast signalling system is, as some like to label it, the most complex thing in the universe.*

The hundred billion neurons are not just stuffed into the skull: they are laid out in columns, bundles, clusters and layers everywhere in the brain, forming the specialized mini-brains that I alluded to earlier. For instance, imagine cutting right into one of the cerebral hemispheres and looking at the surfaces that are exposed as a result. If you were to focus your attention on the cut edge at the very crown of the brain, you would see the separation between the cortex (the "bark" of the brain) and the white matter underneath. The cortex is made up of layers of cell bodies of neurons—it is the so-called grey matter. The white stuff is myelin, a fatty covering insulating the axons of those cortical neurons. The axons are making connections to other neurons, sometimes in distant parts of the

---

* The hundred billion neurons in the brain are actually outnumbered by cells called glia. The exact function of glia is unknown, although so far they have been simply assigned the role of "support" cells. But if they were found to have some signalling role, the complexity of the brain would become super-astronomical.

brain. The myelin insulates the axons, accelerating the nerve impulses that travel along them. Remove the myelin, as happens in multiple sclerosis, and the nerve impulses are disrupted. As dense, complex and vast as the brain is, any neuron can be connected to any other neuron in fewer than seven steps, the brain's version of six degrees of separation.

The cerebral cortex, which usually gets the vote as humankind's most impressive neural achievement, is the convoluted surface that forms at least the upper two-thirds of the brain. It is so highly folded, with some folds doubling back on themselves like an oxbow river, that if it were spread out it would cover, well, depending on which source you believe, an area ranging anywhere from a handkerchief to a tennis court. Okay, the tennis court estimate is ridiculous (although it appeared in *New Scientist* magazine in 1993), but I have seen many other comparisons, including the following: ten inches by ten inches (that's the handkerchief), one and a half square metres (for those of you still working in Imperial measurement, that's a square about forty inches to a side) and the floor of a living room (from a University of Toronto psychologist). The most recent I've come across is the equivalent of four standard sheets of writing paper, which I figure is about a family-sized pizza. Regardless, there's a lot of cortex stuffed in there. It has to be folded like this to fit in the skull.

If you could wander around the cerebral cortex you would see how information in the form of neural traffic moves in the brain. Visual information—an image—is the perfect example. Signals from the eye arrive first at the back of the brain, in what's called the visual cortex. The first processing that goes on here is relatively simple: separating objects from their background and sorting out the orientation, shapes, colours and movement of those objects. After that information is processed, it moves forward in the brain, where more details, such as the actual identity of the object, are extracted. In the case of a familiar face, the early processing establishes that there's an object and that it fits with the known pattern of a face. (A recent estimate put the number of neurons involved in the processing of a single visual image at one million, and that's just in the area devoted to identifying objects. The number of neurons involved in the entire visual brain might amount to a couple of hundred million.) Eventually that particular face (and remember it can be seen at virtually any angle, in any lighting) is categorized as familiar or not, as a personal acquaintance or not. Then the associated details of information and any emotion associated with that face are brought into play. The most important point here is that early processing happens in one place, later processing in another.

That is the grey matter, the stuff of the surface of the cerebral cortex. There are two important organizational principles underlying the structure of the human brain. It *is* modular, with discrete areas having responsibility for specific information-processing tasks. Just as important, however, is the fact that none of these areas is actually isolated from the rest of the brain. The intra-brain traffic represented by the white matter shows that information moves in both directions. Not only does visual information flow forward as more details are extracted from it, but areas farther upstream feed information back to the earlier centres, helping to shape the interpretation.

This is what neuroscientists trying to understand consciousness have to work with: a hundred billion neurons, up to ten thousand synapses each—a mighty flesh-and-blood connection machine. Unfortunately, all brains, even the most accomplished ones, look pretty much the same, giving few clues to their abilities. About ten years ago the director of the Moscow Brain Institute, Oleg Adrianov, admitted that seventy years of analysis of Lenin's brain had revealed "nothing sensational," although he maintained that Lenin's grey matter was "undoubtedly the brain of a talented man." How exactly he could have judged that from the brain alone (if he did) isn't clear.

Dr. Sandra Witelson at McMaster University in Hamilton has examined pieces of Albert Einstein's brain closely,[5] and while she found it to be no different from a set of control brains in many respects, including weight and most dimensions, she did find one thing that might have set Einstein's brain apart from anyone else's: his left and right parietal lobes, the parts of the cerebral cortex that cover the crown of the brain, were wider than normal and the whole brain, as a result, more spherical than most. So what? This unusual size of the parietal lobes together with some unusual topography in those regions suggested to Dr. Witelson that they had developed unusually early, and might have provided the substrate for Einstein's well-known ability to think spatially, visually and mathematically. The parietal lobes are known to be involved in such reasoning. It's hard to be certain of such conclusions when you know you're looking at Einstein's brain—there's nothing blind about the study—but Dr. Witelson did include thirty-five males and fifty-six female brains as controls.

An earlier study of the microstructure of a piece of Einstein's brain also reached the conclusion that it was unusual, citing an unusually high ratio of glial support cells to neurons, and concluding from this that his neurons were working so hard that they needed extra support. Although this study appears in introductory psychology texts, it is practically worthless, partly because the control brains against which Einstein's was compared are so poorly characterized that they are virtually useless for comparison, partly because the statistical analysis was geared to search for a significant result until one was found.[6] I'm sure no one would have predicted that Einstein's brain would have contained more support cells than the cells that actually do the mental work!

It's a tantalizing connection: Einstein's mind was certainly unique, and it would be fascinating to tie that to some feature of the physical brain. But of course the risk is that you would look for such a feature until you found it, and such links have not generally been proven. So we're back to this question: is our current understanding of the brain sufficient to pursue the mystery of consciousness? In the end, that's really up to you to decide, but even those who believe the brain is adequate to the task have to admit that there is a significant sticking point: the mind—your collection of thoughts, emotions, imaginings, memories, images, tumbling together in the stream of consciousness—is an immaterial thing. These features of the mind exist nowhere else but in the mind: they can't be cultured in a Petri dish or captured on film. They have no substance. But the brain is a substantial thing. How then do the two interact? Take the most elementary example: you decide you've had enough of all this and close the book. That action begins with your mind, but eventually sets muscles in motion. Between is the set of neurons that fires—first in your brain and then at your muscles—to cause the action. Ultimately, that is what needs to be explained.

CHAPTER FOUR

The Trickle of Consciousness

FOR consciousness scientists, the challenge is to put two complex things—consciousness and the brain—together. The difficulty is figuring out how the latter gives rise to the former, acknowledging at the same time that there are experts out there who don't think the mind has *anything* to do with the brain.* But if you're buying into the scientific approach to consciousness, then the question "How do you make consciousness in the brain?" becomes *the* question, or, as philosopher David Chalmers christened it, the Hard Problem. Even the "easy" problems are, by Chalmers's admission, hard. In his view these are things like understanding how the brain coordinates sensory information to control behaviour or how we're able to verbalize what we're feeling: challenging, state-of-the-art science. But it all comes down to the hard one: on the one hand you have neurons; on the other, memories and feelings and inspiration, appearing one after the other in rapid succession in your mind. The two have to be connected.

It is a hard problem for good reasons. The brain is a solid, physical object: we can dissect it after death, we can trace its activity electrically or chemically and we can stimulate it and get a reaction.

---

* Obviously, there are as many definitions of "mind" out there as there are people willing to commit to one. But here I'm using "mind" to mean consciousness. Unfortunately, that implies that when you lose consciousness, you lose your mind, and I doubt that my mind ever goes away completely, even when I'm asleep.

There's no reason to think that there's anything going on in there that defies the laws of physics and chemistry. The mind, on the other hand, we're not so sure about. It has no substance, yet it has qualities that substantial things don't have. And you can't even prove that it is real in the same way that the brain is.

For some this difference is a showstopper. The brain can be completely described in terms of its atoms, molecules, forces and energy, and anything a physical system like this does must be explainable in physical terms. If the brain does something, the cause of that something must be physical. But if the brain is doing something, such as lifting your hand because you have willed it to do so, it's acting under the influence of something *non*physical. How could the ghostly mind reach across that impassable barrier and tweak the brain?

Reaching for an analogy, philosopher Daniel Dennett once related the story of how children, when they see the insides of an electronic toy, decide that the computer chip can't have anything to do with what the toy does and that the battery must be the key. He then argued that we're just as boggled by the brain. "When we look at a human brain and try to think of it as the seat of all that mental activity, we see it as something that is just as incomprehensible as the microchip is to the child when she considers it to be the seat of all the fascinating activity that she knows so well as the behaviour of the simple toy."[1]

"Magical" would be a good word to describe it. There have been few attempts to link the two directly. One of the most notable was by Sir John Eccles, an Australian researcher who won the Nobel Prize for his research into the workings of synapses. In his later years, Eccles became as much philosopher as physiologist, in that he sought a mechanism that would allow the immaterial mind to communicate with the material brain without violating the laws of physics.

Eccles turned to quantum mechanics for the answer.[2] In the unpredictable world of the quantum, in which strange events like matter appearing out of nowhere—literally—or like particles of

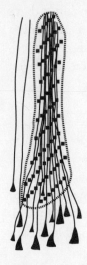

The late Sir John Eccles was one of very few scientists who tried to reconcile an immaterial mind with the material brain. This shows how his "psychons" (the dotted lines)—the elements of mind—interact with the clusters of neurons he called "dendrons."

light instantly transporting themselves from one side of a room to the other are commonplace, it is conceivable that events on a very small scale can occur without the usually required input or expense of energy. Eccles was specific about how this would happen: the sacs of neurotransmitters that are released from a neuron upon the arrival of an electrical impulse are arranged in a gridlike structure, like ice cubes in a tray—it's called a presynaptic vesicular grid. Eccles suggested that while a mental event, such as the urge to do something, could not propel those neurotransmitter sacs to discharge their contents into the space between cells, it could alter the probability that a nerve impulse would cause that to happen. (Many times an impulse fails to dislodge the neurotransmitters).

Ultimately he identified the specific players in this scheme: bundles of neuron endings—the dendrites—clustered together to create what he called "dendrons." Their numbers are staggering: the number of synapses, actual connections between neurons, in a dendron might be over 100,000, and there would be 40 million dendrons in the brain. In Eccles's view they represented the brain side of the equation. Each dendron was then linked to a psychon, which Eccles defined as a mental event or *experience*. The diagrams in Eccles's scientific papers show the dendrons as an inverted cone of dendrites, over which are draped, veil-like, the transparent psychons. The connection between the two is a probabilistic quantum one.

This would make sense in that probability plays a hugely important role in quantum mechanics. Objects are neither here nor there (they can even be one *thing* or another) but behave in a probabilistic sense: they *might be* here or there. In roughly the same way, Eccles suggested, a mental event could tweak the release of neurotransmitters, as long as they and the structures surrounding them were small enough to be part of the quantum world. The beauty of the idea was that there would be no violations of the conservations of energy or of the barrier between the physical and the nonphysical. He put it this way: "According to the hypothesis the presynaptic vesicular grid provides the *chance* for the mental intention to change by *choice* the probability of synaptic emission." For it to work reliably, the mental intention would be focused on many synapses, not just a few. I'm sure most neuroscientists today would view Eccles's scheme as a desperate attempt to reconcile the mind-brain separation without having to abandon the scientific approach that he obviously held so dear. I'm equally sure no one today believes that Eccles was right.

Eccles is one of the few accomplished scientists who has attempted to bring the mind and brain together. Why aren't there more? Many argue that it will all come out in the wash, that once we understand how the brain and its neurons are generating consciousness, it will become obvious how the mind is part of all that. Still others believe that we already have evidence (apart from the questionable John Lorber stories) that the mind doesn't *need* the brain.

One of the most intriguing pieces of evidence came from a Dutch research team in 2001.[3] They scrutinized the accounts of near-death experiences (NDEs) provided by more than three hundred people whose hearts had stopped but who were then resuscitated in hospital. Most of this report was a fairly prosaic account of who had a near-death experience, who didn't, who remembered them, who didn't and so on. But there was one fascinating point raised by the researchers: how could patients remember near-death experiences—with their typical scenes of travelling down a tunnel, meeting deceased loved ones and seeing celestial landscapes—when, if their cardiac arrest was typical, their brains would have been flatlined at the time? The Dutch researchers claimed that such memories couldn't have related to the period just before arrest, because patients are alert at that point. Nor could they have come from the period immediately after resuscitation, because patients then are too confused. That only leaves the time, sometimes minutes, when their hearts had stopped—and typically, when the heart is stopped, the brain is inactive. On those occasions when a person suffers cardiac arrest while hooked up to an EEG, brain activity ceases within ten to twenty seconds after the heart stops and doesn't resume until the heart starts beating on its own again. But if there is no brain activity, how could the scenes of near-death experience be produced—unless, of course, the mind that produced them had no need of an active brain?

The Dutch team put it directly: "The thus far assumed, but never proven, concept that consciousness and memories are localised in the brain should be discussed. How could a clear consciousness outside one's body be experienced at the moment that the brain no longer functions during a period of clinical death with flat EEG?" Then, if that weren't clear enough, they added, "NDE pushes at the limits of medical ideas about the range of human consciousness and the mind-brain relation."

As I think I've made clear by now, that is pretty hard to swallow for most scientists, who are sold on the idea that the brain makes the mind. In fact, a commentary by Christopher French in the same issue of the medical journal *The Lancet* points out that a significant number of patients in the study, when interviewed two years later, remembered having had an NDE even though they hadn't reported one immediately after their heart attack. Theirs are clearly false memories of the event, and if they were "remembered" much later, who's to say that some of the others weren't falsely remembered in the days and weeks following the patients' hospitalization? There is also no actual proof that these memories didn't come from just before or especially just after cardiac arrest—this is just considered unlikely by the authors.*

There *are* ways out of this box. Some, especially David Chalmers, suspect that consciousness will eventually be recognized as a fundamental entity in the universe, like the electrical charge of subatomic particles—it won't be possible to break it down into more basic units. But there are lots of scientists who don't feel that that sort of futurizing is necessary, and instead have chosen at least to pursue consciousness with the faith that once caught, the mystery will unravel. If you are going to take that route it helps to know exactly what consciousness is—or isn't. It turns out that "isn't" is easier to establish than "is." I am willing to bet that by the time you reach the end of this chapter, you'll be thinking differently about your own consciousness.

Without going into detail again, the starting point here is that consciousness is "on" all the time: it makes you aware of your surroundings; it's vivid, all-encompassing and responsible for your thinking.

But how true to life is that sense that you are aware of your surroundings? Not very. There's a tried-and-true trick of demonstrating the selectivity of awareness by asking people, as I would ask

---

* There is no doubt that the opinions expressed by the Dutch researchers are repellent to many, especially hardline skeptics. Michael Shermer, who writes the "Skeptic" column in *Scientific American* magazine, refers to this study in the February 2003 issue only to say that "because the everyday occurrence is of stimuli coming from the outside, when a part of the brain abnormally generates these illusions, another part of the brain interprets them as external events." No mention of the awkward fact that the brain that is generating "these illusions" was almost certainly flatlined at the time.

you now, to pay attention to the pressure of the chair under your butt.* Most of the time that's not hard to do, and most of the time no one is actually doing it until asked. This example shows nicely how a steady stream of sensory information must have been coming into the brain from pressure sensors in the skin, but until you were prompted, none of that information entered your consciousness.

Psychologist Susan Blackmore has made the same point about sounds. If you are at a cocktail party and amid the din of conversation you hear your name, you are often able to reconstruct the sentence or two before, to tune in retrospectively to a conversation that, had your name not appeared in it, would have passed you by completely unnoticed. What does that say about the conversation? Was it in or out of your consciousness? Blackmore also cites the chiming clock whose bells you only become aware of partway through their sequence, but as soon as you do, you "remember" the first few and are able to pick up the count accurately. Were those first few chimes in or out of the stream of consciousness?

The same sort of thing happens in vision. How many times have you run around the house looking for a set of keys or an important piece of paper, only to discover that it had been right on the table in front of you the entire time? In situations like that, I've discovered that the only efficient way to search for something is to prime my attention by visualizing the object, then search around in the real world for a match.

Here's another example: imagine you're in a train station, walking against the flow of traffic as hundreds of commuters push their way by, scanning their faces for someone you know. Now imagine you're in exactly the same station, again with passengers streaming by, only this time you're counting the number of people with freckles. In both cases you'd be looking directly at each face, but you'd be paying attention to completely different things and so would be "seeing" the faces differently.**

---

* My wife hates that word, but I used "buttocks" recently in a newspaper column and people laughed at me for being prudish.

**Psychologist Allison Sekuler pointed out another beautiful example: the arrow between the "E" and the "x" in the FedEx logo. If you've never noticed it before, check it out. It's stunningly obvious.

These examples, drawn from three different senses, illustrate that much of what is apprehended by our senses goes unnoticed. At least when it comes to being conscious of incoming information, the stream of consciousness is more like a rivulet, or even a trickle.

But how much exactly do we miss? It's tricky to calculate how little of what's out there actually contributes to what's in here, but one estimate is that *one millionth* of the information that bombards our senses actually makes it into consciousness. One millionth. How much are those sensory organs transmitting to the brain? Again, although it's difficult to be precise, when measured in bits of information,* numbers like ten million bits a second from the eye, a million per second from the skin and hundreds of thousands per second from the other organs are reasonable. But the amount that trickles through to consciousness is something like forty bits per second. If you do the math, you'll find that these numbers don't add up: forty bits a second is perhaps three or four times more than a millionth of the incoming information, but hey, why quibble? The fact is that consciousness extracts an extremely small fraction of all the information that is available to it.

I'm not saying that it's a bad thing that we only allow the tiniest of fragments of sensory information into our consciousness. Most of it is irrelevant and we would likely be overwhelmed by it, a kind of ongoing bad trip, the sensory equivalent of the difficulties endured by the famous Russian mnemonist Shereshevsky, whose memory was so encyclopedic that he had great difficulty erasing those memories he was no longer interested in.

How much does this say about consciousness itself? Yes, the vast majority of incoming information never reaches consciousness, but consciousness doesn't seem to be the worse off for it. It still has all those qualities of vividness and imagination that make our inner mental lives what they are. Again, however, we

---

* A bit in binary code is the information represented by a yes/no decision, a single on/off switch. A byte is eight bits; a megabyte, eight million bits.

may be convincing ourselves that it is much more profoundly complex than it really is.

For instance, how good are we at juggling things in our minds? How many facts can we keep there in the focus of consciousness? Not many.

Psychologist George Miller established this unequivocally in a paper he wrote in the 1950s, a report rated as one of the most influential ever to appear in the journal that published it.[4] Miller was talking about the amount of information we can deal with at any moment, such as being able to discriminate among tones of different pitch, or the saltiness of different solutions. In surveying a variety of studies, Miller concluded that we are capable of handling about seven different tones, or seven solutions, or seven anything. He called it the "span of absolute judgment." However, he pointed out, we have ways of overcoming this limitation, one of which is "chunking" the information. For instance, typical telephone numbers are seven digits, and if you have to move from one room to another while keeping a number in your mind, you'll likely agree that seven is just about at the mental limit. But what happens when you add a three-digit area code? Most of the time, not much, because area codes, although three digits, can usually be remembered as a single chunk of memory—416 for Toronto, 604 for Vancouver—and so don't put a huge strain on your mind. For that matter, the first three digits of telephone numbers are common enough that many of *them* can be remembered as single chunks.

Although Miller was describing the limits to what we'd call short-term memory, in many ways that brief memory store, also called working memory, *is* the stuff of consciousness. It's what is in your head right now. Since Miller wrote his classic report in 1956, there has been much research on the matter.* If anything,

---

* For that matter, there was a lot of it *before* Miller too. One investigator named Jevons, published a report in *Nature* in 1871 showing that three or four beans thrown into a box could be counted accurately and instantaneously but that six were guessed correctly only 120 times out of 147 throws.

some psychologists today, based on forty years' worth of research since Miller's paper, are willing to lower that magic number to four, or, to paraphrase Miller, four plus or minus one.

There have also been brain-imaging studies that suggest, at least indirectly and with respect to vision alone, that four might be closer to the actual magic number than Miller's original seven. In one study, people who were lying in an MRI were briefly shown sets of objects, ranging anywhere from one to eight in number, then one second later were asked how many there had been. The split in results hovered around four: any fewer than that and they had no trouble remembering; more than four and their accuracy fell off abruptly. In addition, an area of the brain called the posterior parietal cortex, at the upper back of the head, increased its activity up to four objects and then levelled off. That could be interpreted to mean that four was about all the parietal cortex could handle. Interestingly, when the subjects were simply looking at objects without trying to remember them, the parietal cortex responded in the same way no matter how many objects, but visual regions at the back of the brain continued to "count." The easiest explanation for the differences would be to say that the parietal cortex is concerned with being aware of the number, while the other vision centres, which lie upstream in terms of neural processing, are simply doing the routine accounting of what's before your eyes.

This isn't exactly a direct commentary on consciousness, because you can have material in your short-term memory that you are not directly conscious of at this moment. In fact, one consciousness researcher, Bernard Baars, argues that the capacity for items in consciousness is not seven, not four, but one.[5]

Baars allows that this one thing can be a chunk of information, but says that studies of ambiguous figures (such as illusions in which you see either a rabbit's head or a duck) suggest that one thing is all our consciousness can handle. And it's exactly why when you're driving along an unfamiliar street looking for a particular address, you're better off to shut off the radio, because it is the source of a competing stream of information, and unless you can shut it out of your consciousness completely, it'll reduce your capacity to search for the house number.

Think of *two* telephone numbers, one that is familiar and the other a number you just pick at random out of the phone book. It feels to me that I can be conscious of—I can keep in mind—the entirety of familiar phone numbers, all seven digits of them. They are there in my consciousness simultaneously, likely because their familiarity allows me to treat them as a single chunk of information. But unfamiliar numbers are different: as I try to visualize or be conscious of them, I find that I am reading off the digits to myself, two or at most three digits at a time.

Baars goes on to comment that it is extraordinary that, with a brain the size of ours, we can keep so little uppermost in our minds at any moment. We're feeble: a cheap calculator, one that you probably wouldn't bother to stoop down and pick out of the gutter, can remember more numbers than you can. It doesn't make any sense. Wouldn't it be better to maintain as big a store of incoming sensory information and memory as possible, so that we could deal quickly with anything that came up?

It might be that such broadband awareness would be overwhelming, impossible to organize efficiently and dysfunctional as a result. On the other hand, there is some reason to believe that while vast arrays of neurons assembled together as memory storage devices is likely a good thing, there should also be a way of searching through them or of choosing the most relevant ones to apply to any novel situation.

The honeybee is a relatively simple example. Its brain contains fewer than a million neurons and occupies a space of about one cubic millimetre. That is 100,000 times fewer neurons than we have packed into a space 1,500,000 times smaller. Yet honeybees are capable of solving very complex problems. If, for instance, they are trained to fly home in one direction from a feeder in the morning and a different direction from an afternoon feeder but then are placed at the morning feeder in the afternoon, they will nonetheless fly straight back to the hive. They have somehow learned the pattern in a deeper way than simply by rote. In the same way, if bees are trained to fly in two different directions based on two different orientations of a striped pattern, if shown a new pattern that is midway between the two they will fly in a direction that is also midway.

The structure of the bee's simple brain suggests that while there are neurons that are specifically designed to produce a single, invariable response, there are also other neuron modules that seem capable of integrating their information to produce novel responses. What isn't yet known is how that selection of appropriate neurons (or packages of neurons) to make the right response is made.

The honeybee brain might be on the track to an organization that is like many versions of artificial intelligence, and also, perhaps, like the human brain: made up of many different informational modules packaged together with an additional narrow focus mechanism to select the relevant pieces of information when faced with a problem.

But what part of a setup like that would consciousness be? Remember, the brain's overall capacity is vast, and it is processing huge amounts of information at any one time. In the technical jargon, it's a parallel processor, its prowess a result not of the speed of its neurons, but of their number and their simultaneity. But consciousness, by contrast, is linear, a thin line of thought. How does it emerge from the rest of the brain, and how much of the rest of the brain's activity does it capture? The answers to those questions might surprise you.

CHAPTER FIVE

The Unconscious

Civilization advances by extending the number of important operations which we can perform without thinking about them.

—**Alfred North Whitehead, <u>An Introduction to Mathematics</u>, 1911**

THE trickle of consciousness runs counter to the fact that we have an enormously large brain and are capable of very complex behaviour. How is that brain being used? Where does all that behaviour come from? There really is no puzzle. The brain is in use, and it is guiding that behaviour—it's just that we are not aware of most of the processing going on in our brains: it is largely unconscious.

I am not talking here about Freud's unconscious, which stored away unpleasant repressed ideas that only occasionally bubbled up into consciousness in some weirdly disguised form, nor about the collective unconscious that is supposed to be shared by all of us and through which one could trace human history. This unconscious refers simply to the workings of your brain of which you are unaware. Think back to when you learned to drive: every single step had to be considered. Put the key in, turn it *that* way, take off the parking brake, look carefully at the automatic transmission to be sure that you're about to shift into drive and not reverse or neutral, check both mirrors—have you done everything that you should? ease your foot off the brake—it was endless and harrowing. And that was before you even pulled out into the road! After driving for a few weeks or months, the process was totally different. You still did each of these things; to an outside observer the only apparent difference would be that you did them faster, but the real difference was inside: you no longer paid conscious attention to them. Consciousness was always available to be recruited again if something out-of-the-ordinary happened: if you became aware of a car looming in the rear-view mirror, you would snap to attention. But you were also capable of driving—of making the serial decisions necessary to keep the car on the road—without consciousness.

William James captured the unconsciousness of decisions, at least in the days before widespread central heating, this way:

We know what it is to get out of bed on a freezing morning in a room without a fire, and how the very vital principle within us protests against the ordeal. Probably most persons have lain on certain mornings for an hour at a time unable to brace themselves to the resolve. We think how late we shall be, how the duties of the day will suffer; we say, "I *must* get up, this is ignominious," etc.; but still the warm couch feels too delicious, the cold outside too cruel, and resolution faints away and postpones itself again and again just as it seemed on the verge of bursting the resistance and passing over into the decisive act. Now how do we *ever* get up under such circumstances? If I may generalize from my own experience, we more often than not get up without any struggle or decision at all. We suddenly find that we *have* got up.[1]

Most people are surprised and skeptical that things happen that way when I tell them that story, but try it yourself. If you're thinking about the decision-making process, then of course you'll be able to convince yourself that getting out of bed was a conscious decision, but if, as is usually the case, your mind is elsewhere, it happens outside of consciousness somewhere.

Another example, this time supplied by the late Julian Jaynes.[2] Put any two different objects—a coffee cup and a glass, a pen and a pencil—on the table in front of you. Then slowly and deliberately pick each up in turn and judge which is heavier. Concentrate! You will be conscious of all sorts of little details of the objects—their texture, heft, any irregularities in the shape—but are you conscious of the judging process? No. As Jaynes said, "It is somehow just given to you by your nervous system."

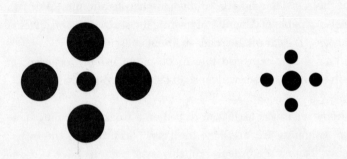

Believe it or not, the two circles in the middle of each set are exactly the same size. Your conscious mind cannot be persuaded that this is true, but your unconscious is fully aware of the fact.

There are hundreds of experiments that testify to the importance of the unconscious in daily life, but a small sample will give you the flavour. Mel Goodale at the University of Western Ontario has demonstrated one of the most striking, an experiment that shows that our movements can be controlled unconsciously.[3] There are many visual illusions that distort the apparent size of an image. One is the Ebbinghaus illusion (see illustration above), in which a single circle is surrounded by a set of circles of different size. The circle in the middle will look bigger or smaller depending on the size of those in the surrounding set. The illusion is so compelling that even when you know that two circles are the same size, you can't escape the conviction—

driven by the illusion—that they are different. Goodale created a beautiful elaboration of the illusion by building a tactile version out of actual discs, such as poker chips. Then he was able to show that even when people are completely convinced that two central discs are different sizes, when asked to reach out and grasp them they calibrate the gap between thumb and forefinger to the exact size of the disc. This is just one piece of evidence amassed by Goodale and David Milner that we have *two* visual systems in our brains—one for identifying objects by their three-dimensional characteristics, the other for actually reaching out and picking them up. That latter system is unconscious and unencumbered by the need to relate the size of an object to everything surrounding it. This means that the reaching system won't be fooled by illusions like this one. Of course, you feel no conflict because the reaching system remains forever out of reach of your conscious mind.

There is also evidence that we can *perceive* things unconsciously, in itself a challenging idea because most of us, and the dictionaries we might consult, see perception and consciousness as tightly linked. Dictionary definitions of *perceive* like "to attain awareness" or "to become aware of something through the senses" are common. But reality is different. Psychologists such as Phil Merikle at the University of Waterloo have assembled the evidence that we perceive things unconsciously, and it is convincing.[4] That evidence reaches back to the nineteenth century. Psychologist Boris Sidis reported in 1898 experiments in which he showed volunteers cards with a single letter or number on them. He held the cards far enough away that no one could make the cards out clearly, reporting only that they were able to see a dim spot or nothing at all. But when Sidis than forced them to "guess" what had been on the cards, they were correct about whether it had been a letter or a number, and even about the identity of each, at a better-than-chance level. Somehow the information had entered their brains without their being aware of it. A possible weakness of this study is the assumption that when people say they can't see the letter or the number, they truly aren't aware of it. You and I would likely accept that testimony at face value, but psychologists have to be careful that they are not assuming too much in doing so. Maybe the subjects' verbal descriptions of what was in their minds wasn't entirely accurate or complete. However, there is plenty of research since Sidis's experiment that has supported his claims.

Throughout the 1980s and 1990s there was a wealth of experiments that demonstrated how pervasive the influence of unconscious mental processing is on our consciousness. The ones that appeal most to me are those which, in a single stroke, established a profound effect. One such experiment was reported by Robert Zajonc and William Kunst-Wilson in 1980.[5] Zajonc and Kunst-Wilson began with the observation that familiarity with things can increase our fondness for them, and cited the case of a piece of music that with repeated listenings becomes better liked (as opposed to "familiarity breeds contempt"). But they then took this perfectly ordinary example into much stranger territory. They had their subjects view a set of irregular octagons: simple black shapes on a white background. The trick was that the shapes were visible for a mere one one-thousandth of a second, and each subject saw only half of the complete set. Then, in the second phase of the experiment, subjects viewed pairs of octagons presented simultaneously. In each pair, one had been seen before, one had not. Subjects were asked to pick out the one they had seen before, and to choose which one they preferred. Now you might think it's silly to be asked to pick one eight-sided figure over another because you like it better, but there was method in this madness. The judgments had confidence ratings associated with them: 3, "sure"; 2, "half-sure"; and 1, "guess."

The results were curious to say the least. When it came to recognizing previously seen shapes, the subjects did no better than chance. That isn't really surprising given that the shapes had been visible for only a millisecond, far less than necessary to be aware of them. But even so, those same subjects preferred the shape that they had seen at a statistically significant rate. Out of the twenty-four subjects, while only five apparently recognized the shape they'd seen in the first round, sixteen preferred that shape. Somehow the subliminal perception of the shape was influencing their judgment of it even though they didn't remember seeing it.

The confidence ratings were pretty weird too: they revealed that at some level, the subjects seemed to have some notion of what was going on in their brains, even if they weren't technically *aware*. When they rated their own judgments as "guesses," they were right: they scored at a chance level in both recognition and preference. But when they thought they were "half-sure" or "sure," they preferred the shapes they'd already seen, even though they still didn't recognize them!

Zajonc and Kunst-Wilson had transported the well-known effect of liking something better as you get to know it from the conscious mind into the unconscious, a remarkable and startling discovery, especially when you reflect on the fact that this effect must be operating on all of us, all the time—outside of our awareness.

At about the same time as Zajonc and Kunst-Wilson were showing that unconscious familiarity affects our feelings about things even if they are nothing more than random geometric figures, Anthony Marcel at Cambridge University was uncovering some equally peculiar qualities of the unconscious mind. Zajonc and Kunst-Wilson had prevented their subjects from being aware of the shapes they were looking at by presenting them for only a tiny fraction of a second; Marcel used a different technique called "masking."[6]

Masking involves flashing some sort of stimulus, in this case a word, followed immediately by a second image, in this case composed of parts of letters arranged randomly. The mask prevents the word from entering consciousness: subjects are simply unaware of what it was. But of course, as we've seen, unaware doesn't mean it isn't there in their brains somewhere. The words for this experiment were chosen to be either graphically similar, that is, similar in appearance, like "acquaintance" and "acquiescence" or "hint" and "hind," or semantically similar, like "acquaintance" and "friend"or "hint" and "clue." Then subjects were asked to match masked words against target words for either their graphic or their semantic similarity.

As expected, Marcel found that the sooner the mask followed on the heels of the test word, the less likely it was that subjects would be able to decide which of two target words it matched. However, he also uncovered something that few had anticipated, and that was that the last judgment to falter was that of the meaning of the word. First to go was the subject's awareness that they had even seen a word (when asked "Is a word present?" they got to the point where they were guessing). But that happened before the subjects lost the ability to match the words graphically, and even when they were in the dark about both the presence of the word and its shape, they were still able to judge whether the word that they were by this time totally unaware of matched the meaning of a target word.

Here is why that result had such an impact. It had been assumed that there was an order to the processing of words, starting with the simple awareness of the word, progressing through its appearance, until after some more advanced processing the meaning became clear. It seems intuitive that the meaning of a word requires more brain power than just making out its outline. On this basis, you might expect that masking the word would cause the loss of meaning first, then the details of its shape, and finally, as the visual information degrades even further, the mere presence of the word. But no, it actually happens in reverse order.

Imagine that: you are presented with a word, you have no idea that you have seen it, you certainly can't remember it, yet you can match the word correctly to a word of similar meaning. You would obviously feel as if you were guessing, and in Marcel's experiment there were three subjects who quit because it made no sense to them to be making judgments about something they had absolutely no clue about. Understandable, but at the point they quit they were still "guessing" correctly between 60 per cent and 70 per cent of the time. Those who shared their feelings that the experiment was "nonsensical" but nonetheless kept going tried adopting different strategies for guessing, but those who decided to let their feelings guide them did the best.

Anyone who had thought that the unconscious mind might simply participate in the early, simple processing of visual images, leaving the complicated stuff to consciousness, must have been shocked by Marcel's experiments. The bottom line is simple: detailed analysis of images is carried out unconsciously, even analysis that would seem to be the kind of thing that only consciousness would be capable of.

It's really impossible to say how much of this is going on even at this moment in your brain. Hopefully you're concentrating on what you're reading, but inevitably there will be images, sounds, touch sensations and likely much more that are sneaking their way into your brain but not into your mind (your *conscious* mind). Those thoughts are there, they are influencing how you think, but you are aware of none of them.

Much of this might come as a surprise to you, but there is an aspect of unconscious processing that we are all familiar with, even if we haven't devoted much thought to it. That is the "aha" experience.

The aha experience, which is unfortunately rare for most of us, occurs when the solution to a problem apparently pops up out of thin air. We have been thinking about the problem, there's no doubt about that, but the solution is suddenly just there. This is obviously intriguing to psychologists, because it suggests that the unconscious has been active again: from where else would a solution suddenly come?

Most psychologists would call this "insight," the whatever-it-is that allows you to solve a problem that you are unfamiliar with, and there have been entire books written about it. It is a mysterious process, with many theories attempting to explain how it comes about, but what most people seem to agree on is that there is a significant unconscious component. The great mathematician

Henri Poincaré argued that while it was important to have a pre-
liminary phase in which you analyzed the problem from every
conceivable angle, the step that intervened after that—but before
the solution—was much harder to define. He called it the "appearance of sudden illumination." Others have referred to an incubation stage that follows the conscious mulling over of the problem.
In fact, some scientists have been specific that the second stage in
reaching an insightful solution to a problem is a solitary one:
going for a walk, having a nap or doing something completely
unrelated to the problem itself. It is during such moments that the
solution to the problem presents itself.

It was while I was revising this chapter that I had an amazing
unconscious/conscious moment. I was flying to Calgary to attend
a meeting and I ran into someone in the Toronto airport who was
going to the same meeting. I knew her, I knew exactly the context
that I knew her in (I had interviewed her years before when I was
hosting the CBC radio program *Quirks and Quarks*), but at this
particular moment I couldn't remember her name. I'm not the
best at remembering people's names anyway. We sat in different
places on the plane and then rented separate cars and drove to
our mutual destination.

But here's how I eventually remembered her name. I thought
about it a couple of times on the plane, and remembered that she
had a double surname, and that her first name started with the
letter K. But that was the absolute best I could do. Then, while I
was driving, I suddenly thought about it again and immediately
had the feeling that I was closer, even though I hadn't been
thinking about it—consciously—for more than an hour. I thought
back to the topic of the interview, the health risks of marijuana
smoking, and her name immediately popped into my mind. My
unconscious mind had done all the work. It was also interesting
that the interview topic triggered her name. I guess they will
always be associated in my mind.

I'm sure the classic story of Archimedes solving the problem of
how to determine if the gold crown given to King Hiero II had

actually been adulterated with silver is another example, but we really don't have much detail about what went through his mind when, having watched the water in the bathtub rise as he lowered himself into it, he realized he could do the same with the crown and determine its purity. This was of course the moment that spurred him to run, naked, through the streets shouting "Eureka!"*
Einstein had a similar moment: "I was sitting in a chair in the patent office at Berne when all of a sudden a thought occurred to me: 'if a person falls freely he will not feel his own weight.' I was startled. This simple thought made a deep impression on me. It impelled me toward a theory of gravitation."[7]

Some psychologists have come up with detailed models of what actually happens in the mind to create such a moment of insight. For example, Janet Davidson and Robert Sternberg have proposed that insight involves three processes of selection: selective encoding, combination and comparison. "Selective encoding" refers to the fact that in solving a problem you may suddenly see something that hasn't caught your attention before, or see that some facts are more relevant to the solution than others; selective combination involves putting disparate elements together in a way you hadn't previously imagined; and selective comparison happens when you're able to put together new information with old in a way that opens the door to a solution.

Some examples make those steps clearer. One puzzle Sternberg and Davidson gave their subjects was this: "One day you decide to visit the zoo. While there, you see a group of giraffes and ostriches. Altogether they have 30 eyes and 44 legs. How many animals are there?"[8] The quickest way to a solution in this case is to realize that the number of legs isn't relevant. If a bunch of two-eyed animals

---

* Most accounts of the insight process in science are careful to point out that the insightful moment must be followed by the arduous and not nearly as exciting process of substantiating the insight. In Archimedes' case, modern analysts doubt that he could possibly have solved the problem the way it is described in the legend; the differences in displacement between a crown of pure gold and one mixed with even substantial amounts of silver would have been too difficult to measure.

have a total of thirty eyes, there are fifteen animals, period. Sternberg and Davidson identify the key process in this case as selective encoding, the art of recognizing the data that suit the solution.

Theirs is not the only theory as to how insight happens, nor do the three steps cover all the bases. For instance, it's generally accepted that one route to insight is to let go of pre-existing ideas, to avoid fixating on what you already know or think. There is a classic example of this from the 1940s.[9] Karl Duncker gave his subjects three cardboard boxes, candles, matches and thumbtacks. The challenge was to figure out how to mount the candles on a door, supposedly for use in experiments on vision. The solution was to melt enough candle wax to be able to attach the candle to the end of the box, then tack the box with the candle on it to the wall. The neat thing was that those who received the candles, matches and tacks *in* the box had a harder time solving the problem than did those who received everything separately. Duncker suggested that the former group had trouble ridding themselves of the notion that the box was to hold things, and so had a harder time thinking of it as a *platform*.

Even the problem of tying the two ropes together I described in Chapter 1, while I cited it as an example of how we can fool ourselves as to what's going on in our minds, is also a perfect example of how, under the right circumstances, insightful solutions to problems happen. And that is really the point: they just happen. Even with all the analysis of insight, even with the volumes of examples of how we arrive at solutions, the bottom line is that consciousness plays no role in the solution, only in the testing of the solution after it appears. Often people who experience insight report that they didn't even know they were still working on the problem. Insight is an unconscious process usually hastened by diverting the conscious mind by the walk in the woods or the long steamy bath, as if continued mulling over the problem by the conscious mind actually interferes with the unconscious mind's ability to get on with it.

Of course, it would be great to know exactly what's going on. How does that unconscious mental processing actually work away at a problem while you are literally not thinking about it?

There have been some recent clues found in brain-imaging experiments, clues that, while not pinning down how the brain creates insight, at least suggest where it happens. Mark Jung-Beeman at Northwestern University and his colleagues posed problems to people whose brain activity was being recorded either by magnetic resonance imaging or EEG (electroencephalography).[10] The problems were like this: subjects would see three words, like "pine," "crab" and "sauce," then try to come up with a single word that would fit with all three, in this case, "apple." At the same time, the problem solvers were asked to say whether the solution, if they came up with it, required insight. This turns out to be relatively easy to do: you know when you've arrived at the solution to a problem out of the blue.

When people reported having used insight, the brain imaging and the electrical recordings of their brain activity revealed specific sites in the brain that had been active, and the two techniques dovetailed nicely. The primary place for insight, at least in these problems that required only verbal problem solving (not conceptual, as in some of the earlier examples in this chapter), was in the right hemisphere, in what's called the anterior superior temporal gyrus, a small ridge at the front of the temporal lobe, just by the right ear. This area was already known to be active when people try to find the theme expressed in a story or to come up with the best ending to an incomplete sentence, so it was no surprise that it might play an important role in language tasks, but the interesting thing here is that it was *not* active when people claimed they had solved the word association test *without* using insight. In other words, this seems to be a part of the brain specifically attuned to insightful solution to tricky word-meaning problems.

The researchers see a link between this part of the temporal lobe and its role in surveying a wide range of words and their meanings, with the suggestion that insight requires a broad focus and linking together of disparate solutions, together with abandoning the old tried-and-true approaches. It was also significant that the electrical recordings revealed a burst of electrical activity in the brain that was about as sudden as the subjective experience of realizing that the problem had been solved. Even more intriguing was that the fact that there was earlier, but much weaker, electrical activity recorded from a part of the brain much closer to the crown of the head. Jung-Beeman and his colleagues suggest this was an unconscious earlier (by one second) "realization" of the solution to the problem, which then gave rise to the more pronounced—and conscious—burst from the temporal lobe.

In the end, unconscious mental processing is most important if it affects our behaviour, and there are experiments that demonstrate exactly that. John Bargh of New York University has produced some of the most imaginative and startling of these.[11] In one experiment, Bargh suspected that stereotypes could modify our behaviour in ways we wouldn't even suspect. One well-established stereotype, especially among college students, is of old age. In particular, there is a common expectation that seniors are slower and weaker than people younger than them. Bargh took advantage of this stereotype in the following way: in the guise of giving volunteers a scrambled sentence test, Bargh exposed half the group to a set of words that included several that were suggestive of seniors, like "Florida," "bingo" and "forgetful," while others saw words like "California," "awkward" and "apples." The experimenters were careful to exclude words directly associated with slowness or weakness. When the "sentence" test was finished, the participating students were led to believe that the experiment was over, and were allowed to leave the building. But unbeknownst to them, a member of the experimental team timed them as they made their way down the hall to a piece of carpet tape at the forty-foot mark. Why? Bargh and his team had predicted that those who had been primed with words suggesting age would take on some aspects of the stereotype, like weakness and slowness, and would actually take longer on average to reach the marker than the control group did. Bargh was right: they did take longer. Although it sounds unbelievable, the only sensible explanation is that these students had apparently internalized the idea of slowness or weakness even though, of course, they had no clue that they had done so. Think about that: not only was the stereotype operating unseen and unfelt, but it actually had effects, not just on how readily, for instance, the participants might leap to their feet to offer a chair to an older person, but on their behaviour unprompted by the presence of anyone else. The experimenters were careful to exclude the possibility that the unusual slowness of the experimental subjects might have been due to their being depressed (undoubtedly at the thought of Florida and bingo together) by testing their mood. If anything, the slow walkers were cheerier than the others!

Bargh and his team performed another clever experiment in which people were shown a videotape of an apparent job interview but the viewers were divided into two groups. One group was told that the job being discussed was that of an investigative reporter, the other that it was a waiter position in a restaurant. The conversation between the applicant and the interviewer was sufficiently general to be able to apply to both jobs, but participants were nonetheless supposed to be evaluating the interviewee's qualifications for the job.

About halfway into the videotaped interview, a third person, named Mike, knocked on the door, entered the room and asked the interviewer if he could come out for lunch. The interviewer said he was too busy, and it would have to wait until later or another time. At this point, what had been exactly the same tape took two different courses. In one, Mike became irritated and claimed that he was just as busy and couldn't wait; the interviewer said he just couldn't leave right then, so Mike bolted. The other ending was very different: Mike apologized for interrupting and said meekly that he would wait outside until the interviewer was ready.

After the tape finished, the subjects were asked—perhaps to their surprise—to evaluate Mike's personality, something they couldn't have planned on because his entrance was unanticipated. Bargh had predicted that Mike's personality would be evaluated differently depending on which job interview subjects had been led to believe they were watching, and again he was right. Those subjects who thought they were watching an interview for the position of investigative reporter liked "Surly Mike" better than "Polite Mike," while those who thought they were watching an interview for a job as waiter liked Polite Mike better. In both cases, the judgment of Mike—who, after all, had absolutely nothing to do with the job interview—was based on the kind of job being discussed. Surly Mike apparently seemed to fit better with investigative reporting, Polite Mike with waiting on tables.*

The funny thing was that even those who liked Surly Mike better admitted that he was rude and disagreeable. They weren't in denial—they just felt, unconsciously, that he fit better with the situation. In alternative scenarios where the experimental subjects focused on Mike from the beginning, he was unanimously disliked. I don't think you should find it that hard to extrapolate this experiment to your own daily life.

I could continue listing experiments for many more pages, but the message should be clear already. The incubation process in mathematics has three steps: conscious question, unconscious working on it, conscious solution. Bernard Baars, whose global workspace theory is a popular explanation for the design and layout of consciousness, argues that you can feel the same three steps in even straightforward questions, such as, "What is your mother's maiden name?" You hear the question, there's a *very* brief pause, and the answer comes up out of somewhere. But when you become aware of it, it's already there, waiting. The scanning of the memory banks and the selection of the name (and whatever had to happen between those two things) were unconscious.

When you put it all together, most of what goes on in our brains is unconscious. Just assemble all the experiments referred to in this chapter, apply them to the much wider range of events and behaviours in a typical day that they might reasonably influence, and you'll have accounted for much of human life. So much for consciousness!

---

* Except, of course, that we've all been to restaurants where Surly Mike was our waiter.

CHAPTER SIX

The Grand Illusion

W HILE it is true that consciousness contains very little of the total amount of information assailing our senses, it is also true that what's left constitutes a major part of the consciousness we have. Of the senses, vision is all-important: we are visual creatures, with something like 50 per cent of our brains devoted to processing and analyzing visual data. That is one reason why vision has been the preferred target of experiments designed to figure out just how consciousness works. Ironically, at the same time the study of vision has made the fragile and limited nature of consciousness even clearer.

To appreciate why some researchers feel that studies of vision have forever altered the concept of consciousness, it is important to start with what vision is not. It is not like a camera. It is true that there are some superficial similarities—there is a lens in the eye, and a light-sensitive lining at the back of the eye (the retina) plays the role of film, tape or, in the case of digital cameras, chip—but beyond these parallel structures, there is very little the two share. The camera captures whatever is in its view, and as long as it is focused and the flash goes off, you have a record of what it was pointed at. The eye isn't like that at all, and one of the best ways of illustrating some of the crucial differences is to look at a handful of dramatic cases of the restoration of sight to those who have been blind since an early age.

Probably the most famous of these dates all the way back to 1728. While this wasn't the first instance of a blind person gaining sight, it was the best documented to that date. The patient was a thirteen-year-old boy who had had cataracts blocking his vision since birth. When surgeon William Cheselden removed the cataracts, the boy began to have some very strange experiences, most of which were nothing like vision as we know it. Most striking was the link between touch and vision. The boy had to relate his novel visual life to his previous experiences by touching the same objects. For instance, Cheselden described how the boy learned to distinguish the cat from the dog by sight: "Having often forgotten which was the cat, and which the dog, he was ashamed to ask; but catching the cat, which he knew by feeling, he was observed to look at her steadfastly, and then, setting her down, said, So, puss, I shall know you another time."[1] Paintings were a particular puzzle:

We thought he soon knew what pictures represented, which were shewed to him, but we found afterwards we were mistaken; for about two months after he was couched [treated], he discovered at once they represented solid bodies, when to that time he considered them only as partly-coloured planes, or surfaces diversified with variety of paint; but even then he was no less surprised, expecting the pictures would feel like the things they represented, and was amazed when he found those parts, which by their light and shadow now appeared round and uneven, felt only flat like the rest, and asked, which was the lying sense, feeling or seeing?

We would never, as the boy did, look at a landscape and think of it as nothing more than a flat surface covered by blotches of paint, but neither would we expect the depth we perceive in that paint-ing to be available to the touch, as he did. When he was shown a picture of his father in a locket, he was astonished that such a large face could be made to fit into such a small space. But aside from these hesitations, the boy apparently adapted beautifully to his new sight, and said that "every new object was a new delight." That is the reaction that most sighted people would expect, but it doesn't always turn out that way.

More recently, Richard Gregory, one of the great vision scientists of our time, studied with a man who had recovered his sight after having lost most of it at the age of about ten months,[2] but this story did not turn out so well. Before the operation to remove corneas that had become opaque through infection, the man, called S.B., could barely make out light from dark in one eye, and only "vague hand movements close to his face" in the other. He was also able to distinguish the colours black, white and red. But this limited vision played little or no role in his life: he functioned as if completely blind.

At the relatively advanced age of fifty-two, he received corneal grafts, and was able, for the first time in fifty-plus years, to see. But it turned out, just as in the Cheselden case 250 years earlier, that seeing wasn't as straightforward as you might have imagined. Gregory and his research assistant, Jean Wallace, first met S.B. forty-eight days after the first corneal transplant. As you might have expected from the Cheselden case, S.B. had not totally adjusted to being able to see—in fact, it was evident that he couldn't see at all in the way that normally sighted people can. His visual behaviour was odd: he would sit still, not scanning around the room as most of us might, but instead paying attention only to an object his attention was drawn to, and then studying it attentively.

On the other hand, S.B. was able to identify objects that he couldn't have ever seen before, but apparently he was able to do it by correlating the touch impressions he had gathered over the years with his new-found sight. For instance, Gregory was astonished that S.B. could tell the time displayed on a clock on the wall in the room where they were sitting. It turned out that he was able to do so because he had always carried a large watch with the crystal removed so that he could gently touch the hands and tell the time. When he regained his sight he was able to translate that felt hand position to what he saw.

S.B. provided an even more dramatic demonstration of his ability to transform the feeling of something into its appearance when he was able to name the magazine *Everybody's*, which the two researchers had brought with them. It turned out that the only letters he was able to read were the first two, E and V; he extrapolated from the two to reason that the magazine was *Everybody's*. But the intriguing thing was that when he had been at the school for the blind he had learned letters by feeling wooden block capitals—no lowercase—and the only two letters in the title of the magazine that were capitals, and so recognizable, were the first two: *EVerybody's*.

But transforming the sense of touch into sight also had its problems. S.B. was good at estimating the length of buses, but not their height, at least partly because he had been able to walk along them but had had no way of reaching their roofs. And when touch was no help, he was often confused: he had great trouble recognizing faces, and had no clue about facial expressions. Simple geometric figures threw him. For instance, he couldn't see that the famous Necker cube was a three-dimensional diagram at all (see page 96).

And he at first thought he could reach down with his hands and touch the ground from a window thirty metres above the ground, an impression that he immediately recognized as ridiculous when his vantage point was reversed and he viewed the window from the ground. Later, Gregory took S.B. to London but found that he was almost totally uninterested in the sights, the only exception being things that moved, like the pigeons in Trafalgar Square.

Sadly, S.B.'s new sight, something most of us would assume to be a great gift, turned out to be substantially less than that. Months after his operations, he seemed disappointed that his life had not been transformed. In fact, he continued to live as if he were still blind, sometimes not even bothering to turn on the light at night, hardly ever watching movies or television. He changed from an energetic and enthusiastic person into a dispirited and discouraged one. S.B. died in 1960, seven years after his vision had been restored.

A more recent case involved a man called M.M., who had lost one eye totally at the age of three and a half and had had the cornea of his good eye ruined by chemical and heat damage. At the age of forty-three, he had a successful corneal transplant for that eye, and so was in much the same position as the patients who had preceded him. However, M.M. was at least in the position of being able to furnish more useful information to those studying him, because by this time, in the early 2000s, magnetic resonance imaging was available to see what was happening in M.M.'s brain as he began to see. And that brain activity proved the point that seeing has as much to do with the brain as with the eyes—or more.

It was clear that although M.M.'s eye seemed to be functioning properly, his brain was failing to respond to the images that were being presented to him. He couldn't figure out what two overlapping transparent squares were; like S.B., he couldn't get the Necker cube at all; and, again like S.B., faces were hugely difficult for him. M.M. had great trouble identifying whether a face was male or female, happy or sad. MRI images of his brain while he looked at faces showed that the area of the brain devoted to faces, the fusiform face area on the right side of the brain, was virtually inactive. On the other hand, the primary visual areas of his brain were lighting up: the contrast between the two areas meant he was seeing the faces, but he was not analyzing them.

On the other hand, M.M.'s motion vision was pretty good, and the only reasonable explanation for this seemed to be that of the many attributes of the visual image, motion is established very early in life—in M.M.'s case, early enough that the appropriate brain connections for motion perception had been established before he lost his vision. Even so, seeing motion was at first difficult for him. When he had been blind, he had skied at an expert level (with a guide talking him down the slope). At first, after the operation, seeing the slope rushing towards him was frightening. Since then he has trained himself to see the shape of the hill better, and he now skis, eyes wide open, with more confidence. However, things are far from perfect. As he puts it, "The difference between today and over 2 years ago is that I can better guess at what I am seeing. What is the same is that I am still guessing."[3]

What these sometimes tragic cases illustrate is that vision is not simply a reflection of the world around us. It is more of a production than a reproduction, an act of creation based on images captured by the retina but not inclusive of all of them, and not limited by them. By the time that this information has been processed, analyzed and edited, and, most important, by the time it reaches consciousness, it has been changed in profound ways. If this weren't true, these people who had their vision restored would immediately experience the same visual world as anyone with undamaged vision. But they don't. They were blind during the precious early years when their brains were building up their capability for vision. If that opportunity is missed, it is apparently lost forever, and these individuals can't perform some of the most fundamental acts of vision that we take for granted, like recognizing the expression on a face. You have to know that vision is like this, active not passive, to appreciate some of the most startling experiments of the last ten years, experiments that erode the role of consciousness even further.

First, a simple demonstration. Take the face cards out of a deck and hold one at arm's length out to one side while you stare straight ahead. Now gradually swing your arm forward, keeping your gaze fixed straight ahead, bringing the card closer and closer to the point at which you're staring, until you can identify it. You're unusual if you can do this before the card is a few centimetres (or, to use the proper measure, degrees) off centre. Outside of a narrow cylinder of clarity directly in your line of vision (a cylinder that is probably only one or two degrees across, roughly the size of your thumbnail held at arm's length), you have only the vaguest idea of what can be seen around you.* It is a flashlight beam of vision. Anything slightly outside that area lacks essential detail; the further off-line, the less detail you see. Contrast that with the feeling that you have as you lift your eyes from this book and look around you: everything in your field of view *seems* crystal clear. But that is an illusion.

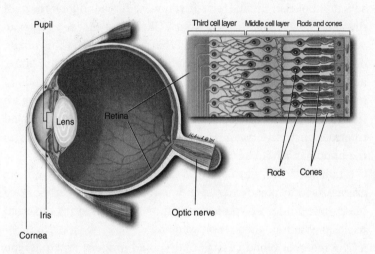

Pupil
Third cell layer   Middle cell layer   Rods and cones
Lens
Retina
Rods   Cones
Iris
Optic nerve
Cornea

The photoreceptor cells of the retina—the cells that actually respond to light—sit at the back of the retina, behind both the cables of neurons that convey that image to the optic nerve and the blood vessels that nourish them. You literally have to see past them. What's even more dysfunctional is that these hundreds of thousands of neurons are bundled together and pass through a hole in the retina to become the optic nerve. That hole becomes your blind spot.

Before we deal with the roots of that illusion, it's important to realize that things are actually worse than this relative lack of detail, because there is a hole in our vision just to one side of that

---

* In his book *Consciousness Explained*, Daniel Dennett created the memorable scene of a room wallpapered with hundreds of Andy Warhol's image of Marilyn Monroe. Upon entering such a room, a brief glance would convince us that the entire room was covered with Marilyns, but, Dennett argues, we would reach that conclusion by telling ourselves it was so, not by actually determining it by seeing them all.

beam of greatest resolution. It is called the blind spot, and it is an absolute blank in the visual scene about the size of a lemon held at arm's length. The blind spot exists by virtue of the fact that the eye is constructed in a clumsy, backward sort of way. The photoreceptor cells of the retina—the cells that actually respond to light—sit at the back of the retina, behind both the cables of neurons that convey that image to the optic nerve and the blood vessels that nourish them. It's as if you were forced to look through the maze of wires connecting your VCR, DVD and video games to see the TV screen behind them. What's worse is that this tangle of wires and plumbing must somehow make its way to the brain, and the most direct route is *through* the retina. So hundreds of thousands of neurons are bundled together and pass through a hole in the retina to become the optic nerve, conveying visual information to the brain. But you can't have photoreceptors in the middle of a hole, and thus the blind spot is born.

If that's the case, why don't you experience the world as a scene crisscrossed by blood vessels and marred by the appearance of a black hole wherever you looked? Your brain performs different tricks to eliminate those two problems.

The mesh of blood vessels that should obscure anything you look at is dealt with by a simple mechanism: anything that doesn't appear to move across the retina disappears. When your eyes perform their habitual, jerky shifts (called saccades) from one feature of a scene to another, those blood vessels, because they're part of the eye, don't shift the way the visual scene in front of them does, and so you can't see them. My favourite demonstration of this principle that "if something doesn't move, it disappears" uses a beautifully designed piece of equipment: a contact lens with a tiny projector mounted on it. The whole assembly weighs a quarter of a gram, and when it projects an image to the eye, that image is stationary on the retina: any movement of the eye simply shifts the lens and its projector with it. The result is that images remain visible for only a few seconds then fade, giving way to a featureless grey, sometimes jet black background. Sometimes they disappear completely, only to reappear later. Sometimes, especially with more complex images, such as the profile of a face, parts disappear one at a time, like Alice's Cheshire Cat.

You can experience something similar on a moonlit night, when the light levels are too dim to register on your colour-detecting cones, the most light-sensitive of the receptors in your eye. The part of the retina that sees the most detail, the fovea, has the highest concentration of cones, and while in normal daylight it is advantageous to orient your eye so that the image you're most interested in falls on the fovea, if you do the same thing in moonlight and keep your gaze fixed, you'll experience the very weird effect that whatever you're looking at gradually disappears. The dim light effectively immobilizes the scene on which the fovea is fixed. The fovea becomes a blind spot.

These built-in mechanisms that allow you to ignore—or be unaware of—the flaws in the design of the eye make good biological sense, and while they don't suggest in and of themselves that vision is incomplete, they do reinforce the idea that it is a complex matter of presenting the raw material from the eye to the brain, which then transforms that information into "seeing."

But strange things can happen during that transformation, and some of those strange things have been uncovered by experiments over the last ten years, experiments whose results are so surprising that they have persuaded some experts that vision's capabilities are so overrated that vision itself can actually be dismissed as a "grand illusion."

One of these discoveries is inattentional blindness. Until you experience it, it is hard to believe, but countless experiments have shown that inattentional blindness is not rare; it is a standard feature of vision. The most famous example of it occurred in an experiment by psychologists Daniel Simons and Christopher Chabris in the mid-1990s.[4] Simons and Chabris asked a group of experimental subjects to view a videotape of two teams of three people passing a basketball among themselves. They were to pay attention either to the three wearing black shirts or to the three wearing white, and to count to themselves either the number of passes or, in a slightly more difficult version, the number of bounce passes and of aerial passes. Partway through, either a woman with an umbrella or a woman in a gorilla suit walked through the group of basketball players. In each case, the intruder was clearly visible in the scene for a full five seconds. After the tape was over, the subjects who had been viewing it were asked, "While you were doing the counting, did you notice anything unusual on the video?", "Did you notice anything other than the six players?" and, best of all, "Did you see a gorilla (a woman carrying an umbrella) walk across the screen?"

The results varied depending on whether people were engaged in the more difficult counting task or whether they were watching the white or black team, but roughly 50 per cent failed to notice either the umbrella-carrying woman or the gorilla, an absolutely astounding result. In a further elaboration of the study, the gorilla walked into the group of players, paused, turned to the camera and beat its chest, then walked away, taking nine seconds for the entire act. Same result: half the group of viewers somehow missed the gorilla. In follow-up questions, Simons and Chabris found that those who had failed to notice either the umbrella woman or the gorilla couldn't recall seeing either even when told they had been there, and many demanded to see the videotape again because they just couldn't believe it. This experiment sounds unbelievable, especially when you see stills taken from the tape, but the results could be, and have been, replicated again and again.

There are other examples that have been conducted with simple letters and figures in the lab, and they suggest, perhaps unexpectedly, that the more similar the novel objects are to those that are being attended to, the more likely they'll be spotted. It's as if our attention, once focused, is unable to stray far from its assignment. (That conclusion is backed up by the fact that in the gorilla study, more participants noticed the black gorilla if they were watching the *black* team.) Even though these experiments have

Stills from the famous "Gorillas in our Midst" video experiment by Christopher Chabris and Daniel Simons. Fully half the people who were watching the white team in the video pass the basketball around failed to notice this gorilla walk through the scene! This is a classic demonstration of inattentional blindness.

taken place in the carefully controlled conditions of the psychology lab, you can't escape the feeling that these findings go way beyond tests that are simply statistically significant or that produce variations from chance. These are real, sizeable and dramatic examples of how we can overlook what should be the most important and salient features of the visual scene because we are simply not looking for them—we are paying attention to something else. How many times might this be happening in everyday life? Your first reaction is to say, "Never," but obviously, based on these experiments, such testimony can't be believed. Maybe a gorilla walked in front of you today and you missed it.

Psychologists were obviously fascinated by these experiments, partly because they led them into interesting paradoxes. For instance, it has been argued that you don't "see" something, in the sense of being aware of it as your eyes fall upon it, unless you direct your attention to it (you don't see the gorilla because you're paying attention instead to the movements of the basketball). But how can you direct your attention to something you're not already aware of? Of course, this runs directly counter to our impression that we *are* seeing everything that passes in front of our eyes. Well, they both can't be right, and it's pretty clear that it's that impression of total visual awareness that has to be abandoned. This is exactly the sort of dilemma that has persuaded some to argue that the sense we have that we're seeing all there is to see is an illusion—a "grand" illusion because it is so pervasive and convincing. It's worth noting that this would be an illusion foisted on us by our consciousness.

However, things are actually even worse than that. Inattentional blindness is only part of the problem. There is also a phenomenon called change blindness, which, although given that label recently, is something that has been known by filmmakers for decades. Change blindness differs from inattentional blindness in that what goes unnoticed is an abrupt change in the scene that is being viewed. Remember, the gorilla walked into the scene and walked out—there was nothing abrupt about it. There are more examples of change blindness than there is space to describe them, but what they have in common is that you view a scene, then there's a brief interruption (it can be a blink, one of those fast eye shifts called a saccade, a momentary dip to black or, as is the case here, a turn of the page), then the scene returns, but something significant in that scene has changed and, of course, you don't notice it.

Change blindness is a nicer example of our visual lack of awareness in that you can be told that something has changed, and you can watch the scene change over and over and over and still not see what the change has been. (With inattentional blindness, of course, as soon as you're told about the gorilla—as soon as you direct your attention to the thing you've missed—you'll see it, because attention is what this phenomenon is all about.) It works with still pictures, as seen in this book, but it also works with real-life examples.

It can be as simple as having two people in a photo change heads—without being noticed—but one of the most celebrated was an experiment conducted by Daniel Simons (again) and Daniel Levin. One of them posed as a man carrying a map, looking for directions on a university campus. He would approach a random passer-by and engage him or her in conversation. After about ten or fifteen seconds, two men carrying a door would rudely walk between them. As soon as the door blocked the view of the passer-by, the questioner switched places with one of the men carrying the board (who happened to be the other author of the experiment); that switch took no more than a second. Now the subject was having a conversation with someone who, despite holding the same map, was completely different from the original: taller, wearing different clothing, with a different voice and, obviously, a different face. But at least 50 per cent of the time, the experimental subject—the passer-by—failed to notice there had been a change.*

When Simons and Levin reviewed their data in this experiment, they found that older subjects were less likely to notice the change. They reasoned that older people might make a very general classification of the person asking directions ("Students—they all look the same") and therefore not notice a change, whereas fellow students would notice such a dramatic change in someone of their own social group. So Simons and Levin ran the experiment again with two men wearing construction workers' clothing and found that this time the students were just as bad at noticing the switch, apparently because they generalize about construction workers.

Simons and Levin were somewhat concerned that the sudden and unusual nature of the interruption (two guys walking by with a door) might have corrupted the results: would the same blindness occur if the change were less dramatic? Apparently, yes: they conducted a similar experiment in which students stood and waited while someone behind a counter ducked behind the counter saying, "Let me just get you these forms," but was then replaced by a second person, who stood up with the forms and handed them to the student. Again the switch was missed by most, including four out of six students who had, just an hour before, attended a psychology briefing in which they heard about the original experiment with the men carrying the door.

As I said, film editors used change blindness long before it became part of the psychological parlance. They knew that a loud

---

* D. J. Simons and D.T. Levin, "Failure to Detect Changes to People during a Real-World Interaction," *Psychonomic Bulletin & Review* 5 (1998): 644–49. This is a much-quoted paper, but I have to say, after seeing Simon and Levin standing side by side, they are *not* dramatically dissimilar in appearance. That lessens the impact for me.

Look at the first picture in each pair carefully, then switch to the second and try to identify what's missing. Most people have difficulty identifying just what it is, even though in both cases the missing piece is central to the scene. These are examples of what psychologists call "change blindness."

noise was sufficient to block filmgoers' awareness of an abrupt and often inconsistent change in scene. Even a different point of view can obscure what would be startling differences. Most fans of *The Wizard of Oz* don't realize that when Dorothy meets the Scarecrow and they dance around at the edge of the cornfield, the length of her hair changes from one moment to the next: long, then short, then long, then short. Until you're told it happens you just don't see it. There are too many examples to list, but another is a scene in *My Own Private Idaho* in which River Phoenix and Keanu Reeves are in deep and serious conversation. Unfortunately, in shots of Phoenix he's holding a stick in his hands but shots focused on Reeves show Phoenix to be empty-handed. There's another in *Ace Ventura: When Nature Calls*, where the pieces disappear from a chessboard from one moment to the next. Simons and Levin have taken this to extremes, conducting experiments in which changes occur at every single cut in a video: two women sit in conversation while a scarf appears and disappears, a hand moves from chin to table, the plates change colour—the authors called it a "circus of discontinuity." Still, the changes go largely unnoticed.

We are not all equally change blind. Expertise is a help. One study conducted by German psychologists showed that familiarity with American football—understandably rare in Germany—helps people recognize when things with some football importance are changed in a scene. The experts even notice when players have changed their position on the offensive line from one picture to the next, something that completely escapes those unfamiliar with the game.

Change blindness has revealed another chink in the armour of the idea that we have all-seeing visual awareness. If just a momentary interruption can obscure even the most obvious changes in a visual scene, what does that say about our awareness of that scene? Well, it says that it's pretty feeble. If we had taken in the complete scene in detail, we wouldn't fail to notice the removal of a significant item. But apparently that isn't the case. Even if we took in only the most important features of the scene, leaving aside relatively trivial details like the texture or pattern of the upholstery on the couch in the middle of the room, we *should* notice if the couch is gone. What if we were conscious only of the one or two most obvious things, or even of simply some abstract features of them? There would have to be enough detail—of the right kind—registered in our brains so that when the pictures changed, we would notice what was missing.

Those psychologists willing to take this the furthest argue that in our short-term memory we preserve almost *nothing* of the visual scene we just viewed and in which we were confident we saw everything. That would make vision a very grand illusion indeed. How can anyone get away with even suggesting something so radical, that we actually are taking in virtually nothing of the scene before us? They can because the data suggest it might be true, and they do have a fall-back position: it seems to us that we are aware of everything before us because we can look at it at a moment's notice and reassure ourselves that everything is still there in its place. As psychologist Kevin O'Regan has argued, the visual scene that we think we're keeping in our heads is actually out there; we can only refresh our hopelessly inadequate perceptual memory by glancing at it once again.[5]

O'Regan and philosopher Alva Noë have tried to soften the blow of this seemingly unbelievable argument by pointing out that while we unthinkingly assert that yes, we do see everything around us, and yes, we see it in the kind of detail that Ted Williams saw in high fastballs, we're really just *saying* that. If we think about it for even a few seconds, or do the simple kind of activity that I mentioned at the beginning of this chapter—looking at face cards—we quickly realize that it's not so much a grand illusion as it is a simple sort of folk belief, which a little thinking can easily demolish.* If you want

---

* O'Regan and Noe also take on the apparent paradox that we seem to see something only if we turn our attention to it but that attending to it would appear to be impossible if we haven't already seen it. They argue that we do see those unattended things, but only in our unconscious. That's enough to direct the attentional spotlight to them, but not enough to be aware of them before the spotlight arrives.

to keep believing in your visual prowess, then it's a grand *delusion*, but if you're willing to face the facts, then you'll acknowledge that it simply isn't true. That doesn't, of course, explain why our consciousness seeks to persuade us that it is all-seeing—that's another question entirely.

Conscious versus unconscious rears its head again here. Some of the most recent experiments in change blindness show that many people, even when they fail to notice a dramatic change, such as the disappearance of a piece of clothing or the change in colour of a basketball being held in a person's arms, can nonetheless pick the piece of clothing or the original basketball colouration out of a choice of four, suggesting that even though they miss the change, they still retain some sort of image of the original. The problem is that they are unable to, or just don't, compare the original to the final so that they can detect the change when it happens.

Here's my question. If we see only the most minimal details of the scene before us and our brains fill in the rest, or at least persuade us that we're aware of it all, why do I have so much trouble figuring out what an entire room will look like if it's painted with the colour I'm looking at on a little paint patch? That seems to me to be something consciousness is perfectly equipped to do, but it sure doesn't do it for me.

The Necker cube, one of the most famous "ambiguous" figures. As you stare at it, it will appear to flip back and forth all on its own (from a cube pointing down and to the right to one pointing up and to the left). You experience the changes, but you don't control them. The cool thing about the Necker cube is that it never changes, but your awareness of it does.

THE NECKER CUBE

Visual illusions create confusion in the mind, but some of them might provide an opportunity to clear up some of the confusion about consciousness. The kinds of illusions I'm referring to are called "ambiguous," images that flip back and forth between two alternatives. The simplest, and one of the most familiar, is the Necker cube. A simple outline drawing, the cube sits at just the right angle that it could be tilted either up and to the right, or down and to the left. Stare at a Necker cube, and it will flip from one to the other, whether you will it to or not. You can follow what happens, but it's not easily controllable. There is some research that suggests that the flip from one orientation to the other takes around three seconds, as long as you're not trying to control it, a finding that would certainly interest neuroscientists: what sort of brain mechanism would there be with that period that would switch perceptions? On the other hand, evidence that the picture isn't straightforward is supplied by the fact that you can influence that three-second shift with your mind. William James argued that you could decide which orientation of the Necker cube you could see: "We can make the change from one apparent form to the other by imagining strongly in advance the form we wish to see."[6] He also claimed that while a first look at such an ambiguous figure may require close examination before the alternative view becomes apparent, once you've seen the two you never forget them.

What's going on with the Necker cube? It's worth remembering that at least two individuals who have had their sight restored, S.B. and M.M., were incapable of seeing even the three-dimensionality of it, let alone the flip from one version to the other, so it's clear that the cube taps high-level visual processing in the brain.[*] That processing fascinates psychologists because the two versions of the Necker cube are produced by a single drawing. So there is absolutely no difference in the information being analyzed by the visual system in your brain, but there are two radically different conscious experiences. British visual expert Richard Gregory put it perfectly in his book *The Intelligent Eye*:

---

[*] There is a neat experiment that anyone can do that plays on S.B.'s and M.M.'s ability to transfer their knowledge about touch to their new-found vision. If an actual wire-frame 3-D model of a Necker cube is painted with luminous paint, so that it glows in the dark, then held in a dark room, the image will still flip back and forth, even though the touch information doesn't change. It is, according to those who've done it, a very weird experience.

"Evidently there are two equally probable solutions to the perceptual problem: 'What is the object out there? The brain entertains each of its hypothetical solutions in turn—and never makes up its mind.'"[7]

Illusions like the Necker cube provide a golden opportunity to search for the place in the brain where this flip-flop is happening. If someone in an MRI watches the Necker cube and experiences it changing from one orientation to the other, would that change be visible as a change in activity in different parts of the brain? Because the cube itself doesn't change, any changes in brain activity should represent the experience, the perception of the cube. If that could be done then you would have closed in, if not identified, parts of the brain responsible for consciousness.

Tantalizing, but difficult. The problem with the Necker cube is that it's likely that the brain areas responsible for the two different orientations are close together—too close together to be able to separate in an MRI. However, there is a different illusion, called Rubin's vase, that might get around that problem. In your mind Rubin's vase alternates between an image of two faces and a single vase. From the brain's point of view, faces and objects like vases are analyzed in different places, places sufficiently far apart that brain imaging could distinguish them easily. Several studies of this illusion have shown that there are indeed changes in the principal locations of brain activity that happen exactly at the time a person experiences the flip-flop of the illusion, and also that the brain areas involved are widely separated.*

These experiments make it clear that when you're looking at the Rubin's vase and it suddenly shifts to the two-face version, a different part of your brain has stepped up its activity. Does that explain consciousness? Well, no, but it does illustrate that there are significant changes in brain activity that occur along with a moment of awareness, and that's exciting. It is, however, only true for one small part of the brain, because as you're gazing at the illusion, you're also aware of many other things. The illusion may be the primary thing in your mind, but it's not the only thing. However, it is getting closer to consciousness.

Why do ambiguous figures like the Necker cube and Rubin's vase flip back and forth? What is driving the changes? No one knows that, but there is a way that you can experience a related phenomenon, one that is much used by consciousness researchers, called "binocular rivalry." It results in the

---

* See Chapter 15, page 244.

same thing: two different conscious images derived from a single situation.

Here's how you can experience binocular rivalry: all you need is a toilet paper or paper towel roll and an interesting scene to look at. Hold the roll up to your right eye with your right hand. At the same time, hold your left hand, palm facing you, just touching the cardboard tube, so that you're looking through the tube with your right eye (at some scene with a lot of detail) and directly at your palm with your left. At first your palm will seem to have a hole in it, but with time it will fill in. Then you'll find that the two images alternate, in the same way as the two versions of the Necker cube do. This isn't a perfect version of binocular rivalry, partly because your palm is too close and slightly out of focus, partly because it just isn't as interesting as what you're looking at with your right eye (although if your palm is tattooed that would be pretty cool). But as with the Necker cube, you really can't control the alternation of the two scenes: they are rivals. Brain imaging would likely reveal some significant differences in the parts of your brain that are active at those moments.

The Blind Spot

We should expect, whether using one or both eyes, a black or dark spot upon every landscape within 15 [degrees] of the point which particularly attracts our notice. The Divine Artificer, however, has not left his work thus imperfect. . . . The spot, in place of being black, has always the same colour as the ground.

—Sir David Brewster, 1832

Sometimes the simplest demonstrations require the deepest explanations. The blind spot is one of those. You must have experienced this at some point in your life, most likely in some children's book or magazine. As you can see on page 100, there are two objects involved: in this case just a plain circle and a cross, but there have been more inventive versions. King Charles II of England apparently used to behead—in the virtual sense— any one of his courtiers he chose with his blind spot. But making it fancy misses the point: the intrigue and entertainment lie in the blind spot itself.

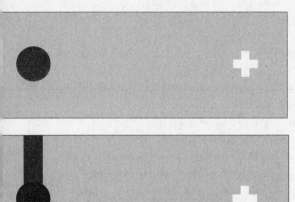

In any of these pictures, you can find your blind spot by closing your right eye and positioning your left directly in front of the cross, focusing on it as you move the page slowly in and out. There will be a place, usually around 20 centimetres from your face, where the circle disappears from sight.

If a vertical line is one shade above the circle, and another below, the filled-in shade is usually indistinguishable. If the line was the same shade throughout, the fill-in would be that shade.

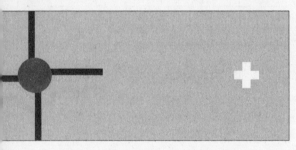

If four lines extending from the circle are misaligned, the two verticals will appear to be straight within the blind spot, but the two horizontal remain misaligned.

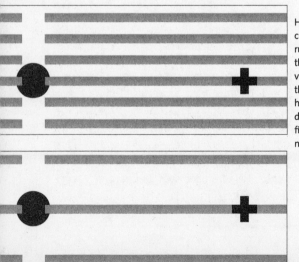

Here the gaps in the horizontal lines create the illusion of a vertical line running down through the array. If there are enough horizontal lines, the virtual vertical is continued through the circle. But if the number of horizontal lines is reduced dramatically, to two or three, then filling in completes the horizontal, not the vertical.

Here's how you experience it. Close your right eye and position your left eye directly in front of the cross, focusing on it as you move the page slowly in and out. There will be a place, usually around 20 centimetres from your face, where the circle disappears from sight. You have found your blind spot.

One in each eye, the blind spot is created where the neurons from the retina—the sheet of light sensitive receptors at the back of the eye—meet to form the optic nerve, which carries the visual information gathered here off to the brain for processing. Each blind spot lies just to the nasal side of the centre of the retina. The photoreceptive retinal sheet is interrupted here, with the result that the visual field of each eye is missing something in the middle. It's a blank space, a surprisingly big one: each blind spot occupies a space equivalent to seeing ten full moons lined up. For the technically minded, it's about 5 degrees across; the full moon is half a degree. To put it another way, it creates a blank equivalent to the size of a lemon held at arm's length.

So why don't we, as David Brewster suggested, notice a big blank in the middle of every visual scene we survey? Partly because the two blind spots don't coincide. Our two eyes cover different (although overlapping) segments of the scene before us: that is crucial for binocular—3-D or stereovision—but it also means that the blind spot of one eye pokes a different hole in that scene from the blind spot in the other. The flip side is that the area erased by one eye's blind spot is seen by the other eye and vice versa. Normally, then, as long as both eyes are open, the entire scene is covered, blind spots notwithstanding. But when you shut one eye, as you do with this demonstration, you eliminate the right-eye coverage of the left eye's blind spot, and with the paper in just the right position, it pops into view.

Or does it? Once you've revealed your blind spot, you will appreciate the second reason we don't see a big blank superimposed on the visual background. There is no blank because the blind spot isn't a hole in vision; it blends into the scene around it, apparently combining texture, pattern and colour to resemble what the eye actually sees.

You need to do some simple experiments to experience just what the blind spot feels like. For my money, the best descriptions belong to Jerome Lettvin, a visual scientist at the Massachusetts Institute of Technology. In 1974 Lettvin wrote an article in which he described some of his own reactions to blind spot experiments.[8] One such experiment involved pushing the point of a pencil slowly through the blind spot. Lettvin wrote,

"For me, the end of the pencil disappears in a curious way. The pencil does not end at the boundary as if it were cut off, but, instead, becomes nonexistent. There is no boundary—in the sense I ordinarily use the term—that marks the apparent end of the pencil. The transition is as if all visual properties vanish. The pencil end becomes nondescript in a nondescript way."

I love that description because it is so personal. If you try a similar exercise yourself, you may have a different reaction, but who cares? Lettvin has highlighted the fact that visual experiences are all different, however much they have in common. He goes on to try orienting the pencil in different ways, then sticking pieces of different coloured or patterned paper into the blind spot to see if anything seems to fill in that spot. If you have the time, it's worth trying such exercises, because what seems to happen to objects that intrude into the blind spot has been an important issue in consciousness research.

On the printed page, the blind spot simply seems to blend in with the rest of the white background. The "lost" area is white, or at least whitish. Is that an issue for the understanding of consciousness? Apparently it is, if the amount written about it is any gauge. Philosopher Daniel Dennett believes this phenomenon has produced some beautiful examples of what he thinks is the wrong approach to consciousness. He is one of those who has campaigned aggressively against the outmoded notion of your brain being a little theatre with your representative sitting in it; there's no representative at all. Nothing has to be "presented," it just is. So Dennett takes issue with anyone who suggests that the blind spot is "filled in" by the brain: painted blue (in whatever way brain cells and their chemicals would do that) if viewed against a blue background, plaid if plaid. He argues that the blind spot doesn't have to be filled in, because there is no place in the brain expecting to get information from that part of the retina. There have never been any light receptors there, so there's no brain tissue that has adapted itself to analyze their output. It's like 108.9 FM: you don't miss it because it's not on the dial.

Dennett's explanation sounds good, but at least some of it is wrong. All kinds of experiments with varying kinds of blind spots have demonstrated that we do indeed fill in the blind spot, although maybe not as exactly and faithfully as you might have thought or as Dennett might have claimed. Neuroscientist V.S. Ramachandran has created a zoo of blind spots that has rewritten the understanding of how they work.

By elaborating the straightforward black-and-white blind spot, Ramachandran was able to demonstrate that colour may or may not be filled in. If a vertical line is drawn through the disk that is to disappear, the line will be completed through the blind spot and the filled-in portion will appear to have the same colour as the line. However, if the line is one colour above the disk and another below it, the filled-in colour is usually indistinguishable. If the line is erased either above or below the disk, it then appears not to impinge on the blind spot.

Ramachandran also created disks with lines extending out in the four directions of the compass. If the north–south lines are longer than the east–west lines, the longer arms appeared to be filled in, but not the shorter ones. If the four lines are misaligned, the two verticals will appear to be straight within the blind spot but the two horizontal lines remain misaligned.

If there were any doubt remaining that filling-in is a complex process, one final example should remove it. This display begins with an array of horizontal lines, with a gap in each one about a third of the way along its length. Those gaps create the illusion of a vertical line running down through the array. The disk is then put in one of those gaps, about in the middle of the array. If there are enough horizontal lines, the resulting effect is that the virtual vertical is continued through the disk. But if the number of horizontal lines is reduced dramatically, to two or three, then filling in completes the horizontal, not the vertical.

With these and other displays, Ramachandran apparently successfully made the point that Dennett was wrong: there is indeed "filling-in" of the blind spot. But this is a slippery business, as you might have found if you have tinkered with your own blind spot.

There are two more things that need to be factored in here. First, while there are no photoreceptors in the blind spot itself, if you were to follow the path of the optic nerve to the main visual area at the back of the brain you would find neurons there whose field of view takes in the area around the blind spot. Even though they have apparently nothing much to do, when an object appears in the blind spot these neurons become active. Are they filling in? It looks like they might be, but no one knows yet for sure.

Second, some recent experiments suggest that what you see in your blind spot may depend on how much attention you are paying to it. If you place a pencil with the blind spot halfway through it so that you see both ends of the pencil on either side, the blind spot seems to change depending on whether you focus your attention directly at it or instead pay attention to the two ends of the pencil. Dragging your attention away from the blind spot itself seems to enhance the apparent filling-in, while focusing directly on it seems to inhibit the process.

Filling in the blind spot is a perfect example of how, while it is true that any scene we "see" is concocted by the brain from visual input, those raw data are embellished, sometimes to an elaborate extent, usually without our knowledge. Of course we say with confidence that "we" are seeing, but we simply don't realize how little of the visual information that is available actually enters our awareness.

CHAPTER SEVEN

Time

THE more we learn about consciousness, the less impressive it seems. It captures only a tiny fraction of the information inbound to our senses every second. Most of the processing and refining that is performed on that trickle of data is unconscious anyway. However, we could still console ourselves that, in the end, the best, the highest-quality, optimally relevant information is the stuff that actually enters our minds, creating our selves and our conscious life. It may not be an exact representation of what's out there, but it is a high-level executive summary, a distillation of the outside world that creates each person's unique inner world. It equips us to think, speak and act. But even that much-reduced picture of consciousness exaggerates its fidelity to the world around us. One intriguing example is the perception of time.

Time is one of those contradictory subjects that is familiar and lacking in mystery yet resists easy explanation. Even physicists, now, thanks to Einstein, comfortable with the idea that the passage of time varies depending on who is measuring it, aren't much help when it comes to characterizing it. It is not a force, not matter, not some kind of mysterious field. Time seems to us to flow in only one direction, from past through present to future, but the laws of physics don't demand that—they work equally well with time moving in either direction. There is even exotic physics that suggests that some sort of limited time travel is possible, although nothing yet imagined comes close to the travels of H.G. Wells's *The Time Machine*, or even *Back to the Future*.

We surround ourselves with timekeeping devices because our feeling of the passage of time is unreliable. For every person who claims to know—without looking—what time it is to within a couple of minutes, or who is able to wake up just before the alarm day after day, there are the rest of us, who are as often as not shocked when we check our watches to realize that we "had no idea what time it is."

There wouldn't be a stream of consciousness in the absence of time. That's where it gets its "flow." The two appear to move forward linked tightly to each other, thought for thought, second by second. You might suppose that it is time that paces consciousness, but the reverse can be true as well. The timekeeping mechanisms in our brains are neural and chemical, not mechanical, and as such are influenced by other neural networks or neighbouring chemistry in the brain. There is no shortage of examples of how our personal timekeeping can be disturbed. The biochemist Hudson Hoagland discovered a simple one in the 1930s: his wife was suffering from a fever, and the hotter she was, the more distorted her sense of time became. At one point he went to the local pharmacy for medication, and although he had been away only twenty minutes, she was sure it had been an hour. Chemical reactions speed up with increasing temperature, and her fever had accelerated her clock.

The flip side of Mrs. Hoagland's experience is the ubiquitous feeling that time passes more quickly as we age, a phenomenon that seems to depend primarily on one of the timekeeping mechanisms in our brains gradually running slower (like so many other things) as we get older. A slower clock allows more events to occur in a given minute, hour or year, and so time seems to be passing more quickly. (There are researchers who contend there is more to it than that, that at least part of the time-is-passing-more-quickly feeling is due to a quickening of the pace of life around us.)

Then there's the watched pot that never boils. We're capable of paying only so much attention to anything, and the more focused we are on external events, the less we reserve for keeping time. Intent on the pot, we lose our grip on the passage of time. When you add to that the common illusion that the outward-bound trip always takes longer than the return, it's clear that, depending on circumstances, our sense of the duration of time passing may *resemble* clock time, but we can't hope for much more than that. These examples don't say anything profound about consciousness because they are intermittent and variable, but they open the door to the idea that the perception of time is more subjective than we feel it is.

But how subjective can it be? Surely if you strip away the effects of circumstance that create illusions of time, there must be a steady timepiece at the heart of it all. After all, consciousness feels as if it's seamless, unbroken and happening right now. You are reading these words . . . *now*. Of course, the "now" immediately slips into the past, but for that split second, it was the current moment. If anything it seems to exist slightly in the future: as I'm writing this, I'm putting together the end of this sentence and thinking about how I'm going to move on to the next one—or not. But every other image and thought of the conscious kind— the light coming through the window, the hum of the computer, the mental reminder of the time I have to leave the house—is embedded firmly in the present. Or so it seems. As long as we focus on relatively large chunks of time, minutes or even handfuls of seconds, this is largely true. But there are strong indications that when it comes to shorter intervals, this seldom-questioned notion is completely wrong, that the timers in your brain that keep pace with consciousness are more about creating illusions of time than keeping it.

One of the most interesting examples is the phenomenon called chronostasis, the stopping of time. There are several ways of inducing it, but the most convenient can be done with a clock or watch with a hand that ticks off seconds, or any kind of digital timer, like a microwave clock. The illusion is set up first by not looking at the microwave, then turning to it to see what time it is. Your initial impression will be that the counter seems to be frozen: it is as if the first tick is delayed. Then, once the ticking resumes, it carries on at the normal rate. Turn away again, look back, and the same thing happens. The first change of the numerals (or in the case of the watch, the first movement of the second hand) takes much longer than the subsequent moves. The beauty of this illusion is that you can abandon it at any time, confident that after a break you will inevitably experience it again as you glance at your watch or the microwave.

Researchers in the UK first examined this weird phenomenon in 2001 and decided that the human brain was responsible for the time delay. The problem it faced was a short blank in its experience, the time taken by the eyes to shift from wherever they were originally focused to the clock. During this movement—a saccade—the eyes have no time to focus on or to register anything in their path. Only when they reach the ultimate target can they do that. Rather than have you perceive that time of movement as a meaningless blur, the brain back-times the first image of the clock to cover the time the eyes were moving.

At least that's what the researchers who first investigated chronostasis thought, but since then the picture has gotten slightly more complicated. For one thing, chronostasis can happen when you are waiting for the phone to be answered. These days you usually don't have to wait too long before voice mail kicks in, but if the person you're calling is technologically challenged, you're stuck there, listening to ring after ring. If during that time you are distracted, either by holding the receiver away from your ear and talking to someone or simply by paying attention to something else, when you return to listening, the silent space before the next ring will seem much longer than any of those that follow. It's a close analogy to the microwave illusion, but it involves sound, not sight, and therefore can't have anything to do with saccades.

In controlled lab conditions it has been possible to show that chronostasis occurs not just with vision or hearing, but with speaking and even touching a keypad. Alan Kingstone and his group at the University of British Columbia have suggested that when you lump all these results together, chronostasis is all about attention. When you shift your eyes to the microwave clock, you are anticipating the first number you see, and that moment when your eye first settles on the number stretches out—only for a fraction of a second, but an easily noticeable fraction. It is a micro version of the watched pot delaying its boil. Kingstone has shown that if volunteers are asked to attend, not to the first number they see but to the fourth, it's the fourth that appears to last longer. Wherever the magnifying glass of attention is moved, the moment of time in its focus is dilated.

What determines how long the illusory pause should be? The early experiments suggested that the longer the saccade, the longer the illusion, because there was a greater time gap to be covered up. But that obviously can't apply to settings where the illusion is auditory or tactile. Nonetheless, it can't be random; Kingstone's group suggests that attention, memory and practice all play a role. They also point out that we are very good at judging time in activities like driving and sports, all of which are performed largely unconsciously. It may be that, once again, allowing consciousness to intrude into an area of mental activity that is usually off-limits leads to peculiar judgments.

The experiments that have been done so far have required the presence of an accurate timepiece to reveal the illusion. But we have to assume that such illusions are happening constantly as we go through the day—at least when we are conscious of time. Without a clock to reveal them, they are no longer illusions, just part of the seamless web of events that our brains creates for us. The day doesn't feel as if it is filled with papered-over blurs of the eye or unduly long pauses before telephone rings, but it is. There is likely no time during the day when our mental timekeeping is free from the influence of attention, no time when it runs at the steady clocklike pace that we are convinced it does.

At least we can experience chronostasis given the right conditions. If we move to even finer time scales of time, we're not aware that anything odd might be happening, but there is clear-cut evidence that invention—or at least clever editing and production techniques—is in play when the brain is marking time. Imagine a simple visual scene: you're out in the park, sitting on a park bench, watching people and their dogs. A tall man in a dark grey suit passes in front of you, walking left to right, and at that same moment he is passed by a young woman wearing red sweatpants and a yellow T-shirt, walking a standard poodle almost exactly the same colour as the man's suit. They pass; you glance at them and move on to other things. A simple, easily decoded scene, right? Well, maybe not.

One of the surprising facts about vision is that the different features of a visual scene are analyzed in the brain separately then, having been processed, are recombined. You are aware only of the recombined product. Those features include colour, form, location, movement and orientation. There is no dispute about this: if a map of the visual cortex at the back of the brain were a country, then separate visual modules would be its provinces, each responsible for a different visual feature. Region V4 is responsible for colour, V5 for movement. They are anatomically distinct, and because of that, they can be differentially injured and incapacitated by strokes or tumours.

Patients who have suffered damage that is restricted to area V4 become blind to colours but suffer no other disadvantage. What is fascinating about these patients is that even though they used to see the world in all its colours, they cannot remember what that was like; vision to them is now, and seems always to have been, shades of grey. Damage to the motion areas, V5, also called MT, is more disruptive. The most famous case was the one I mentioned earlier, of the woman who, having suffered a stroke, experienced the world around her as a series of stills. Such patients provide unambiguous evidence that incoming visual information is segregated into its component parts, processed independently, then only later reassembled.

It is curious enough that things happen this way, but what's really fantastic is that the separate processing steps take different amounts of time. The location of things is processed first, followed by colour, then orientation and motion. Each lags behind the other by a significant amount of time, totalling nearly two-tenths of a second. So the man in the dark grey suit and the woman in the red sweatpants provide a challenge for our consciousness. We have no problem assigning the right colours, shapes and movements to the two humans and the dog, but how exactly do we do that? Are the features that are processed earliest held in waiting for the others, colour momentarily stalled, processed but not yet conscious, in a sort of mental nowhere land until motion can catch up?

Obviously, most of the time we don't encounter—or we over-look—the illusions or mistakes prompted by these time differ-ences in our visual systems. That's just as well, because life could get really confusing. Take the example of an intriguing illusion called the colour phi effect. It's pretty simple. A green dot is flashed on a screen, then disappears, to be replaced almost immediately by a red spot that flashes just to its right. If the timing is right, what you think you've seen is not a green dot appearing then disappear-ing, followed by a red dot—you are sure that you've seen a green dot *move* from left to right, and change colour to red *on its way*. You can find a good presentation of the illusion on the Internet, at http://www.philosophy.uncc.edu/faculty/phi/ Phi_Color2.html.*

The colour phi illusion—or at least part of it—has been a favourite of consciousness experts for years now. The fact that the dot appears to jump from one place to the other is unremarkable; that is how motion pictures work. Your brain is making a sensible (but in this case incorrect) judgment that an object that appears first in one location then another has jumped from point A to

---

* A more commonplace version can be found in traffic lights. When red changes abruptly to green, it may appear as if the light has jumped and changed colour on the way.

point B. That's fine, but the changing colour is a different matter. How is it possible to see a green dot changing colour on its way to the second dot, when the second dot hasn't even appeared yet? Unless you can predict the future, you need that second dot to appear. But if you need to be aware of the second dot for the illusion to work, isn't it already too late? The first dot has winked out by that time. The only reasonable explanation seems to be that your brain pauses and creates the illusion after the fact, convincing you that everything happened in real time. Once it has finished splicing in the bit about the changing colour, it runs the final edited (illusionary) version by you to convince you that a green dot flew by, changing to red on the way.

This illusion is somewhat famous—or infamous—among consciousness researchers because it's become kind of a cause célèbre. The philosopher Daniel Dennett seized on this illusion, in his much-quoted book *Consciousness Explained*, as yet another example of what he thinks is wrong-headed about others' thinking about consciousness.[1]

Dennett argues that there is no need for the brain to take in the information, edit it, then run through the whole reconstructed scenario once again so you can experience it. It's all part of his campaign to rid the world of what he thinks is the worst holdover from old-fashioned thinking about consciousness: the idea that there is a little theatre somewhere in the brain where the stream of your consciousness flows by for you (or some little mini-you) to watch. There can be a theatre, all right, but nothing's playing there; instead, bits and pieces of thoughts and perceptions pop up here and there, sometimes disappearing before you're aware of them, sometimes holding your attention for the few moments that most thoughts do. According to Dennett, they aren't just the raw materials of consciousness—they are consciousness itself.

When it comes to the colour phi illusion, Dennett contends that it is actually impossible to know exactly how our conscious selves are fooled by the illusion. It could be an inaccurate memory of what happened, or an accurate memory of an illusion, something that didn't really happen. But regardless, he argues that your brain sorts it out without having to replay the whole affair after deciding what must have happened.

Once you add the fact that different aspects of any visual image are processed at different times, the colour phi illusion becomes even more remarkable. It's really not just one illusion, but two: the well-known one by which we become convinced that the dot changes colour in mid-jump, and another that somehow disguises the fact that information about the two colours is processed after the information about the two locations. Disentangle all that if you can!

When all this is put together, you can't avoid the conclusion that our *awareness* of time passing is nothing like the independent measurement of that time would be. It could be more, it could be less; it is uncertain, illusionary and, to a large extent, made up. It represents a further erosion of our feeling that what we're experiencing inside is a faithful replica of what's happening outside. But erosion in one thing; destruction is another. And that is what resulted from the celebrated—and criticized—experiments described next.

CHAPTER EIGHT

Free Won't

THE examples in the previous chapter establish that brain timekeeping isn't rigorous, isn't accurate and often just springs from our imagination. But so what? We still are obviously able to live our lives unperturbed by the fact that time isn't steadily ticking but is slowing, speeding up, swelling, contracting and even reversing its field. We're still in charge; past is past, future is future and we still inhabit the present. Or at least we feel as if we do. But that sense of the present is much more profoundly disturbed by a set of experiments now decades old. Even though some of them were first performed in the 1960s, the results remain highly controversial, because the conclusions that could be drawn from them are unnerving (no pun intended) to neuroscientists.

Benjamin Libet, a neurophysiologist in San Francisco, is the designer, promoter and defender of these experiments. He took two completely different approaches to the timing of consciousness: one set of experiments explored the question of when we become aware of sensations arriving at the brain, the other the question of when we make the decision to do something, to act. Put those together and they include most of what you do during the day. Libet found that neither is the way it seems.

His first experimental subjects were people who suffered from a serious neurological problem that required the insertion of electrodes directly into the brain to help alleviate their symptoms. Some of these patients had movements disorders, such as Parkinson's disease, while others had severe intractable pain; in both cases, electrodes had been implanted directly into their brains to deliver pulses of electricity to places where they might alleviate the symptoms. This experimental setup is reminiscent of Wilder Penfield's work in the 1950s: in the run-up to surgery for severe epilepsy, Penfield had to map the brain before cutting into it, to prevent accidentally damaging key brain pathways. In several cases, Penfield exposed the side of the brain, the temporal lobe, and then explored the surface with an electrode, trying to find areas that, if cut, wouldn't deprive the patient of any key intellectual or speech capacity. It was through these explorations that Penfield found that the mere touch of the electrode could evoke strong images, many of which seemed to be recollections of long-lost memories.*

Some of Penfield's claims have been controversial, but none of them stirred the pot as Libet's have. It occurred to Libet, as it had to Penfield, that having that rare access to the living brain opened the door to some tantalizing experimental possibilities. So, in his first set of experiments, Libet explored the timing of touch.[1] Some patients had an electrode planted in the somatosensory cortex, a strip of brain running across the top from one side to the other. It is one of the most important destinations for touch sensations travelling from the body to the brain. It maps those sensations (which otherwise are volleys of nerve impulses indistinguishable from any other) to their place of origin. The impulses reporting that you've stubbed your toe are only recognized as such because they arrive at the toe region of the somatosensory cortex. This area is laid out in topographic fashion, with body parts arranged

---

* At least, it was accepted for years that the fragmentary sounds and visions triggered by the electrode were memories, but it became clear eventually that at least in some cases the patient had never been in the circumstances they "remembered."

in a line from one end to the other.* It then makes sense that an electrode firing a tiny pulse into any particular place on this map should create the feeling that the corresponding body part was being touched.

Once Libet showed that this was indeed the case, he had the potential for an experiment: comparing the patient's awareness of an actual touch on the hand with stimulation of the related area of the brain. Touch a hand first, then compare that to a touch on the part of the brain that signals a touch on the hand. In both cases there was a significant delay before the patient became aware of the sensation. The current passing through the brain electrode could be adjusted so that the time necessary to reach so-called "neuronal adequacy" was approximately the same for both the brain and hand. A typical delay was 500 milliseconds, or half a second. That is, you become aware of a touch on the hand about half a second after the actual touch; your awareness of an apparent touch also follows a half-second after your somatosensory cortex is stimulated. (This doesn't include the actual time of travel from the hand to the brain, which is a short enough interval—10 or 15 milliseconds—to be negligible in these experiments.) The sensations themselves differed too: jolting the somatosensory cortex produced a feeling of a "touch on the hand" that wasn't as sharply on and off, not as precise in location as the real thing, and different in quality. That isn't really surprising, considering that the sensation arose from the completely unnatural source of an electrode planted in the brain. But the delay in attaining awareness was the key.

Having established that, Libet was ready to perform the first of his sensational experiments. He first zapped the patient's brain in a place corresponding to, say, the right hand, waited 200 milliseconds,

---

* Because body parts are laid out in order in the somatosensory cortex, an image of one's body can be re-created from them. But their size is proportional to their touch sensitivity, and the image that results is a disturbingly distorted version of the real body, a creature that should rightfully be haunting some dark fairy tale. It is reminiscent of Golem, and is called a homunculus—just like the lone customer in the theatre of the mind, but this one is real.

then touched the patient's actual left hand. Which hand would the patient report first? That 200-millisecond difference—a fifth of a second—should have given the brain stimulation enough of a head start that the patient would become aware of an apparent right-hand touch first. But that didn't happen. The patients consistently reported that their left hand had been touched first.

How could this be, that the sensation from the left hand actually won the race to awareness? There didn't seem to be any way for it to shortcut the half-second build-up of neural activity: Libet had established years before that the skin touch requires that delay, so he saw no reason to invoke that as a reason for this timing anomaly. In fact, Libet could identify no confounding technical or procedural factor: the results were the results—they needed to be explained—and he finally settled on an explanation that was spectacular, surprising and controversial, and remains so to this day.

He argued that to make sense of the experiment, you had to suppose that the touch on the skin only *seemed* to win the race. It had actually lagged behind the touch on the brain, but the brain then *back-timed* the sensation arriving from the skin so that it appeared to the patient as having occurred at about the same time as it actually happened. In other words, the brain erased the half-second delay. Libet argued that in everyday life, this back-timing fools you into thinking that your awareness of a touch and the actual touch took place at the same time. By contrast, such back-timing would not exist for a direct touch on the brain by an electrode, for the simple reason that nothing in the evolution of the human brain would have required it.

Libet called this temporal sleight of hand "subjective referral of the conscious experience backwards in time." Again, it seemed to be required by the experimental evidence: a stimulus, whether touch or electricity, had to last long enough to give rise to awareness. Too short, and you'd never know it was there. Long enough to create awareness, and you have the problem that it is now half a second out of date. So the brain kicks it into the past to give the sense of being aware *now*. But how does the brain know how far back in time to place the sensation? If it back-timed sensations a full second, you'd be "feeling" things before you actually touched them. If it didn't back-time enough, you'd feel the touch long after it was over.

Libet found what he thought must be the marker to guide the back-timing. When the hand was touched, the first electrical activity triggered (long before awareness) was a spike in the EEG called an evoked potential, or EP. Libet suggested that the backwards timing used that EP as the point to turn the clock back to. This made sense given that when the brain's somatosensory cortex was touched directly with the electrode, no EP was created, which might explain why there is no back-timing in that case. Even to those of us who aren't either neuroscientists or philosophers it seems weird: to experience something as simple as a touch on the back of the hand, one's brain does the double trick of taking time to create the sensation then erasing that time so as to make it appear as if the sensation did truly arise at the time of the touch. But there are plenty of experts who think it's not only weird, but absurd, impossible or just wrong.

The criticism ranges far and wide, and whether you end up siding with or against Libet, one thing is true: the man attracts more critics and rebuffs them more aggressively than anyone else in the consciousness field. Entire issues of scientific journals have been filled with critiques, and Libet's rebuttal targets each and every one for dismissal. Sometimes he even has to rebut rebuttals of his rebuttal, but he does it with apparent relish, and without giving an inch.

It would be tedious to give a lot of space to the critiques (and it's not clear how many consciousness experts are convinced by them), but a sample will give you an idea of which aspects of this experiment concern them most. In one of the original descriptions of these experiments, Libet gave space to psychologist Donald Mackay, who had suggested to Libet in conversation that the patients had experienced the *illusion* that they had felt the sensation earlier than they really had, and that there was no need for a brain mechanism to perform the temporal switch. As we saw in the previous chapter, there are illusions of time, and some of them are very consistent. Why couldn't that be the case here? Libet acknowledged the possibility, but felt that this explanation was "less satisfactory" than his own.

Many criticisms have centred on the (necessarily) small number of patients and the inherent uncertainty of having to depend on their "eyewitness" reports of the order of the touches: were they really sure that the hand sensation came first? Though it may have seemed absolutely conclusive to them that one preceded the other, there is no way that we as outside observers can prove that this is the way it actually was; we are totally dependent on the patient's testimony, flawed though it might be. Daniel Dennett—again—took this question of the validity of the reports one step further, by invoking his argument that when it gets down to timing events in tens or even hundreds of milliseconds, there is no evidence that the absolute order of things is ever represented in the brain.[2]

To an outsider, this open combat is kind of attractive, in the same way that the threat of a street fight attracts a crowd. Emotions run high on this topic. But why? Why are so many critics determined to undermine the conclusions of Libet's experiments, when, at least in this case, he's really only claiming some neural legerdemain to sort out some annoying delays and make the course of events in the world seem right to the conscious person? But it does go way beyond that, and the implications that upset people are seen most easily in Libet's second set of experiments.[3]

These turned the story on its head. They addressed not the sequence of events in feeling something, but the question, "What happens in your brain when you make the decision to do something?" This time, because he didn't need to deliver an electrical stimulus to a place deep in the brain, Libet worked with people who had recording electrodes taped to their heads. They were then asked to make a slight movements of the fingers or wrist *whenever they wanted.*

They had no time limit, no minimum number of such movements to make in the course of a session. Just whenever they felt, they should move. They did have one other obligation: they had to report exactly when they made the decision to move, and this they were able to fix by keeping an eye on a timer with a rotating dot, like a sweep second hand, on a table in front of them.

So the sequence of events from the patient's point of view was make the decision, check out the timer and move the finger or wrist. Pretty simple. Now, if you had to guess what was revealed by the recording electrodes, I'd bet you'd guess that at some point before the actual movement, there would be some electrical activity that signalled the person's intention to make the movement (after all, the decision to move is made inside the brain somewhere), then maybe some more activity as the brain creates the signal to activate the muscles, then, after the actual movement, it should all fade away.

But that isn't what happened. There was indeed a burst of activity in the brain that appeared to represent the decision to move. The problem was, it came *before* the patient reported making that decision. So in a typical case, electrical activity started in the brain, and about three-tenths of a second later the patient reported making the decision to move, and about two-tenths of a second after that, the patient moved a finger.

Is it any wonder such results startled and confused the experts, sending many of them into denial? This experiment was saying that when you make up your mind to do something, it's too late: your mind has already been made up. The activity necessary to move your finger has already started—at an unconscious level— by the time you say to yourself, "I'm going to move my finger now."

The implications certainly didn't escape Libet. He realized that this was, taken to its extreme, an assault on free will, evidence that our conscious minds are mere passengers—or less, onlookers—in the unending series of decisions that our brains make every day. Concerned about the enormity of the situation, Libet made an attempt to save at least some shreds of free will by making the following argument: while it is true that the conscious decision to move a finger arrives late on the scene, there is still some time, about two-tenths of a second, before the action actually takes place, time enough to exercise a "veto." So free will, at least in this scenario, isn't about making the original decision to do something; it's about stopping the countdown when an action is already on the launch pad, a 200-millisecond window for having second thoughts. It's not free will; it's "free won't."

If Libet's experiments are right, then our mental lives are very different from the way we experience them. A typical day seems full of conscious choices—weighing the evidence, calculating the outcomes, then making decisions one way or the other. A typical day may *look* like that, but Libet's experiments suggest that while, yes, it is true that decisions are being made, they're made *before* we become aware of them. Every once in a while we do exert our will to change our minds. Of course, if we do exercise this conscious veto, we then revert back to the unconscious mind to make another choice. As soon as that decision is made, we then become aware of *it*.

If anything, these experiments have brought more criticism raining down on Libet than the previous set. Much of this criticism revolves around the subjectivity of the whole thing: can you rely on experimental subjects' introspection, the process of monitoring and reporting what's going on in their own brains? Should that sort of report carry the same weight as the traces of electrical activity recorded from that person's brain?

Imagine you're taking part in the experiment. You're to make the decision to flick your finger or wrist and at the same time note the position of the hand on the clock. The clock face is marked off in sixty divisions, exactly like the minutes on a normal clock face, but the spot of light near the outer edge makes a complete turn about every two and a half seconds, so it is speeding along. Your accuracy is important, given that the results are measured in milliseconds. You are to stare at the centre of the clock, note the position of the light when you first feel the urge to move, then remember that time for a later report.

It is technically challenging. Although Libet satisfied himself of the consistency and reliability of the task, you wonder how easy it is to identify the first moment of the intention to act—even on its own—without the complication of having to refer to a clock face. (Remember William James's caution on the difficulty of identifying one's own decision making.) Then add the task of locating the exact position on that face of a rapidly moving spot of light. Finally, you must correlate the two to report where the spot was when you made up your mind. None of this happens on the spur of the moment. It takes some time for you to identify the position of the light: its visual image has to pass along the optic nerve, be processed in the visual cortex at the back of the brain and eventually enter your awareness. Then there's the process of tallying the two together. As some critics have suggested, that span of time might just be enough for subjects to forget or misremember just how much of the process they were conscious of, or to be subject to illusions.

However, the doubts expressed over this work are not limited to technical difficulties. Some question the interpretation of the results: perhaps focusing on a single burst of neuronal activity is mistaken; maybe there are traces of activity crucial to the decision that are masked by the overall activity of the brain. Perhaps what people are identifying as "awareness" is actually just the peak of what was a gradually growing urge that actually did start back when the readiness potential first began firing. As one commentator put it, "one dim moment of consciousness lost in the glare of the next few brighter ones." Or maybe because flicking your wrist or your finger is such a simple and unimportant task, consciousness wasn't recruited to play the same role as it would have been had there been more important choices to be made.

Other questions have been raised about the veto, Libet's last-ditch attempt to retain some role for consciousness in the making of decisions. How exactly is the conscious mind able to veto an upcoming act unless it is aware not only of the act, but also of the potential impact of that act? It has to know what the "brain" is up to in order to cast its veto or not. Not only that: if the veto is a conscious act, then it should require the same amount of unconscious warm-up time before reaching consciousness as any other mental act. But there are only 200 milliseconds available for a process that appears to take half again as long.

Some even argue that this experiment doesn't explore free will at all, because the participants know they have to make several of these "spontaneous" movements over the course of the experiment, and that the movement will be exactly the same (or essentially so) each time. In fact, when the experiment was first described to the participants, they were conscious of the instructions and the procedure—somewhere in their minds they "knew" what they were doing. Yet to take the results at face value you'd have to believe that this background awareness was suspended each and every time they flicked their wrist. Some critics argue that this tightly constrained experimental environment is a far cry from the everyday decisions to take the collector lanes on the freeway instead of the express lanes or to use strawberries instead of blueberries in this morning's pancakes.*

It is reminiscent of the dilemma of the particle physicist who,

---

* Daniel Dennett wants to pull the rug out from under the whole show by claiming that there really is no sense in timing any of these events in the brain. Dennett's argument is that to decide that one event happened before another, you need some sort of finishing line, but there isn't one in the brain.

The following passage represents Dennett's views on the moment in Libet's experiment when the participant relates the position of the spot on the clock face to his decision to flick his wrist: "The 'time of occurrence' of the internal representation? Occurrence where? There is essentially continuous representation of the spot (representing it to be in various different positions), in various different parts of the brain, starting at the retina and moving up through the visual system. As the external spot moves, all these representations change in an asynchronous and spatially distributed way. Where does 'it all come together at an instant in consciousness'? *Nowhere*" (D.C. Dennett and M. Kinsbourne, "Time and the Observer: The Where and When of Consciousness in the Brain," *Behavioral and Brain Sciences* 15 [1992]: 183–247).

Dennett's argument depends on how fine the time distinctions are. He allows that in a case like counting out loud—1, 2, 3, 4, 5 . . . —it makes sense that the number 1 enters consciousness before 2, 2 before 3 and so on. But Libet's experiments deal in thousandths of seconds, and Dennett contends that in that realm, things are different.

when she looks as closely as possible at the fine-grain structure of the world, cannot simultaneously locate a subatomic particle and detail its movement. Or is it like relativity, where the time of an occurrence is different depending on where you are and how fast you're travelling? At any rate, according to Dennett (and others), it is impossible to be sure of the sequence of events in the brain when the only source of information about them is the testimony of the participant in the experiment.

I'd be underselling the importance of Libet's work to emphasize only the critiques. There are plenty of researchers who take the results at face value and are trying to take the next step.

Jordan Peterson and his colleagues at the University of Toronto have tried to make sense of Libet's work by arguing that we spend most of our mental life in the future, not the present.[4] Peterson makes two points. The first is that any time you take some familiar action, even something as routine as reaching out to grab your cup of coffee, you are turning on well-practised motor routines: the choice of muscles, the order in which they activate, the feedback from the feel of the handle, the tightening of your grasp on it, the tilting of the cup to your lips are all in a sense replays, actions that have been accomplished countless times in the past. A better example might be a professional pianist, who, as she plays, is paying attention not to the notes she is playing at the moment, but to those she is about to play and to the overall flow of the music. The more accomplished she is, the farther ahead in the score she will be reading. She is conscious not of the present, but of the future.

Peterson's second point is that, while our conscious mind is focused on the future, we feel as if it's paying attention to the present. As far as the conscious mind is concerned, the future *is* the present. Peterson is taking Libet's backwards referral of time, but he uses it to move the future into the present, rather than the present into the past, as Libet did when explaining why sensations appear to be immediate when they are actually delayed by about half a second. Of course, there is a risk in treating the future as the present—it might not turn out the way you anticipate. Peterson acknowledges that in rapidly changing and unpredictable situations, consciousness might not be of much help. That might be why, in such circumstances, we make the wrong choice.

Benjamin Libet's experiments are supremely controversial because they challenge some of the fundamental working assumptions of scientists who investigate consciousness, the most important of which is that consciousness is to be found in the brain and its neurons; it is not something over and above that, and certainly not separate from the brain. A small minority of brain researchers committed to the idea that mind and brain are separate (critics argue that Libet is among them) took these findings to mean that consciousness is not tied directly to the working of the brain. How could it be if the awareness of the intention to do something follows the brain activity that creates that very intention?

Libet himself has been challenged by the results of his experiments. They have forced him to come up with a theory of consciousness much more radical than those of most of his fellow neuroscientists. He thinks there's a "conscious field" called the CMF (the conscious mental field) surrounding the brain and generated by the neurons in it. This would be a field in the sense that it would extend from and around the brain and, like an electromagnetic or magnetic field, be immaterial. But the similarity to electricity or magnetism would end there. The CMF would be, in Libet's words, "not describable in terms of any externally observable physical events. . . . The CMF would be detectable only in terms of subjective experience, accessible only to the individual who has the experience."[5] The field would be intimately connected to the brain's neurons. Because split-brain patients seem to enjoy two separate consciousnesses in right and left hemispheres, Libet argues that the CMF has to be in contact with the brain to influence consciousness—it can't leap across gaps. But how to test this elusive field?

The problem is to see if a piece of brain, detached from all other neurons, could still register consciousness by virtue of connecting with the conscious mental field it is immersed in. Bearing in mind that the field can't leap over tall buildings, that detached piece of brain would have to remain in place in the brain. There is a precedent for severing neural connections but leaving the piece of brain right where it was: the frontal lobotomy. Surgeons performing this operation in the 1950s and 1960s simply inserted a scalpel up through the eye socket, waved it back and forth cutting all connections to the frontal lobe, then withdrew the scalpel.

The procedure that Benjamin Libet envisions for the detection of the CMF is more delicate and precise, but just as strange. He proposes using a fine wire to cut a small cube of brain tissue free from its connections to neurons on all sides and below. It would be crucial to make these cuts in a way that preserved the surface membrane, which contains the blood vessels that nourish the tissue. A number of other conditions would have to be met, the most important of which would be to find a patient who was already scheduled for removal of a piece of scarred cerebral cortex to eliminate epileptic seizures. Such a patient could have this procedure done immediately prior to the main surgery. The patient would have to be awake, to be able to report any conscious experience, but brain surgery has been carried out many times in the past with patients fully awake and verbal.

If all those conditions were met and the piece of brain that was to be isolated happened to be in an area where stimulation reliably produces a conscious experience, the experiment could go forward. (Libet estimates there are at most five to ten such patients each year.) The surgeon, after having isolated the slab of brain, would stimulate it with an electrode, in the same way Wilder Penfield did with his patients back in the 1950s. If the patient suddenly reported feeling, seeing or otherwise being consciously affected by that electrical current, Libet's CMF would suddenly be on the map. The only way an isolated piece of brain could "report" its experiences to those areas responsible for reporting the experience (speech areas in the left hemisphere) would have to be via some extra-brain connection—a conscious mental field.

It's worth remembering that this theory, whether you buy it or not, would never have come into being had Libet's experiments produced unremarkable results. What if the first signs of electrical activity in the brain had happened at exactly the same moments that the subjects identified as the time they had made the decision to move? Nothing remarkable—no need for a theory of conscious fields. But that's how science moves.

CHAPTER NINE

Free Will

B ENJAMIN LIBET's experiments are controversial on many levels, but they are most disturbing because, if substantiated, they appear to provide concrete neurophysiological evidence that we lack free will, that our conscious selves are mere bystanders in the decision making of the day and that the only shred of control we might retain is "free won't." But this is only one set of experiments, representing one tiny corner of the issue of free will. Philosophers have debated the much broader issue of free will versus determinism for millennia without reaching agreement. I'm interested here in one small part of that debate—whether whatever amount of free will we have (if any) is *conscious*—but the classic debate is worth a quick look first.

Free will describes the ability (which we routinely assume is ours) to control our own actions—the capability, whether we use it or not, to decide freely how we are going to act. Not to have free will would leave us not much more than automatons, behaving according to a set of preset controls or routines, and that is simply not the way it is. Or at least, that's not how it *feels*. But as we've already seen, feelings are an unreliable indicator of the way things are, and there has long been an opposing school of thought: determinism. Determinists argue that free will is an illusion, that no matter what we think or feel, our actions and our decisions are preordained. Two hundred years ago, determinists were able to fall back on physics as evidence for their views. Isaac Newton's brainstorming of the laws of motion may have provided the impetus for this opinion, but the most famous statement of this kind was made by French mathematician Pierre Simon de Laplace: "An intelligence knowing all the forces acting in nature at a given instant, as well as the momentary positions of all things in the universe, would be able to comprehend in one single formula the motions of the largest bodies as well as the lightest atoms in the world, provided that its intellect were sufficiently powerful to subject all data to analysis; to it nothing would be uncertain, the future as well as the past would be present to its eyes."[1]

In other words, if you know the way things are right now, down to the smallest detail, you know the way things will be. It might have seemed to nineteenth-century scientists that de Laplace's "intelligence" was at least conceivable (and with the right kind of faith, you could believe it existed), but the discovery that more often than not nature behaves chaotically rather than deterministically ended the fantasy of proving determinism with science. There are countless instances in nature—the weather being the most obvious—where even if the processes involved might function in a rigidly deterministic way at some level, miniscule (to the point of being unmeasurable) differences in starting conditions lead inevitably to huge differences at the end. It's the old idea of the "butterfly effect": a butterfly flapping its wings in a certain way in Malaysia eventually results in a typhoon over Japan. The world and the universe are much more unpredictable than gigantic games of billiards.

However reassured you might be by chaotic nature's refutation of de Laplace, can you be sure you're not some sort of robot with feelings, acting in a preprogrammed way? Think of it this way: determinists could argue that the correct approach would be to trace the cause of an action back to its source. So you might start with the brain—the source of your will, free or not—then work your way down through neurons, neurotransmitters, molecules, atoms and finally subatomic particles, all of which operate according to well-defined physical laws. Admittedly, there are links in this chain that do not operate completely predictably, but there are *constraints*. You inherited a certain kind of brain. To the extent that brain wiring, the numbers of neurons and their ability to form new connections are set by the genes you inherited, your brain structure is at least partly determined. Add to that the well-known influence of early environment on the development of the brain, something that is too late to change by the time you've passed puberty. Even the experiences you have every day make and remake connections between neurons as you learn and remember, and those in turn are at least partly responsible for your decision making and behaviour.

And there's more: brain imaging shows that the talk therapy of psychologists and psychiatrists changes the way the brain works—an MRI will show different patterns of activation after such therapy. Even if the technology of neuroscience isn't yet up to the task of pinning down or demonstrating the organic basis of every aspect of thought, each of us could make very long lists of the influences on decision making that have taken up residence in our brain as a result of experience and education. How about your conscience? Or the miniature devil and angel that sit on your shoulder and try to persuade you to act in their interests? And don't forget that the influences you are able to identify must represent only a small fraction of the total; most of them are certain to reside in the unconscious.

None of this is the granite determinism of de Laplace, but as it adds up, you can see that it's not impossible or out of line to argue that as you read this, your mental hands are tied—loosely, maybe, but enough to question how free your will is. And if we were completely determined by outside agencies, how would we ever know? You can bet that consciousness would do its best to persuade us that we were in control—or rather, that *it* was.

That brings us out of the sometimes murky waters of philosophy into the neuroscience of free will, and, specifically, conscious free will, the idea that we *feel* as if we are consciously making decisions, day in, day out. If you take the time to examine that process in yourself, you'll recognize that it consists of at least two steps: willing something, then actually doing it. The two are normally in lockstep, but if they aren't, if that link can be broken, then one of the most important features of human consciousness—the feeling that we are responsible for our decisions and actions—can be disrupted.

Here's a trivial but familiar example. We all know it's impossible to tickle ourselves. Experiments have shown that the problem is likely that the brain sets up expectations of what we'll feel if we move our own hand to tickle the bottom of our own foot and then proceeds to discount the sensation, so the effect of the tickling is dramatically dampened. However, if someone else tickles us, there is no opportunity for the brain to make any sensible predictions about when and where the sensation might strike, and its ultimate impact is much, much bigger.

That much is intriguing but fairly straightforward. It's either you doing the tickling or someone else. But what if the distinction between the two isn't so obvious? One of the experiments that examined self-tickling blurred that distinction by setting up an apparatus that allowed people to tickle themselves *indirectly*: a robotic arm actually delivered the tickle (a foam pad dragged across the palm of the right hand), but it in turn was controlled by the subject's left hand. So the subjects were tickling themselves, but through an intermediary. The trick came when the timing of the tickle was changed. When the robotic arm delayed the actual tickle by a moment or two after the subject's hand had moved it, the sensation was more ticklish—the longer the delay, the greater the ticklishness.

Why? Because even though it was apparent to the subjects that they were tickling themselves, they were (we are) accustomed to their actions leading directly to consequences. If you're tickling yourself, the tickling action is followed immediately by the sensation. But if the tickling action is separated from the sensation by as little as two-tenths of a second, doubt creeps in as to whether you were actually the agent of the action. The tickle feels more like it was delivered by someone else; the brain has no chance to link action and sensation, and so it is more ticklish.

That is a small but significant example of how intention and action can be separated and how that separation can weaken the conviction that we controlled the action. Dan Wegner, a psychologist at Harvard University, has explored this more than any other, and has concluded that as long as a thought and an action share three important qualities, the action will seem as if it was willed by us, whether it was or not. Those three qualities are priority, consistency and exclusivity, and together they make up what Wegner calls the Theory of Apparent Mental Causation—emphasis on "apparent."

By "priority," Wegner means that the thought must occur just before the action that it apparently caused. If you flick a light switch and the power goes off at exactly the same time, you can easily convince yourself that you were responsible. However, having the power go off just before you flick the switch, or thirty seconds later, wouldn't have the same effect. Wegner and University of Virginia psychologist Thalia Wheatley demonstrated the effect of priority vividly in an experiment suggested by the age-old mystery of the ouija board.[2] In their case, the ouija was a computer mouse fitted with a small platform on top, just large enough that two people could perch their fingers on it and move the mouse. One of the two people was in on the experiment; the other was the experimental subject. They sat together in front of a computer screen on which was displayed a typical scene from the *I Spy* books (the ones where you have to find a series of objects embedded in a mess of others).

The experiment consisted of a series of steps. The subjects in the experiment were told that they would have about thirty seconds to move the mouse in slow deliberate loops around the screen, then music would start to play and, a few moments after that, the mouse should be stopped. They were then to indicate on a sheet of paper how much of a role they felt they had played in determining *where* the mouse stopped, on a scale of 1 to 10 from "I allowed the stop to happen" to "I intended to make the stop."

The key to the experiment was that the subject also heard an occasional word in the headphones. Some of the words referred to objects that were portrayed on the screen; some didn't. They were told that the words were simply there to distract them, and that what they were hearing was different from what their experimental partner was hearing (the one who was part of the plot). In fact, the partner heard all the same words. This provided a perfect opportunity to test whether subjects could be convinced that they were controlling the mouse, even if they weren't.

In some trials, the partner was instructed to move the cursor to rest on or near the object referred to by the word the subject had heard. It soon became clear that timing was everything. Subjects who heard the word "swan" thirty seconds before the cursor stopped on the swan were unlikely to believe they had had anything to do with it. But if they heard "swan" five seconds before or, even better, just one second before the mouse stopped, they were usually convinced they had willed that stoppage, even though the person across the table was actually the one responsible for moving it there. On the other hand, hearing the word a second *after* the mouse stopped created no sense of having put the mouse there.

Wegner and Wheatley interpreted the experiment as showing that we are likely to believe we have caused something to happen as long as our actions precede the event by a reasonable span of time: one or five seconds before, not thirty seconds before or one second after. (The experiment also addressed the issue of consistency: when you have your fingers on a computer mouse and the cursor controlled by that mouse lands on a picture that represents a word that is already in your short-term memory, the possibility that you caused the mouse to arrive there is consistent with the situation.)

If nothing else, this experiment demonstrated that we can easily persuade ourselves that we are responsible for events around us even if we're not, provided circumstances allow for the possibility of that responsibility. Wegner noted that this experiment was inspired by the ouija board experience, during which the participants do will the planchette (the miniature three-legged stool) to move but are unaware of their role, and usually refuse to acknowledge it. Applying Wegner's three conditions—priority, consistency and exclusivity—makes sense of the ouija experience: when the planchette moves about the board, the movement is likely to be inconsistent with what was just going on in your mind, so priority will likely be rare; what was going on in your mind is also unlikely to tally exactly with what the ouija is saying, so consistency is a problem; and there is at least one other person—usually several—involved, so you aren't acting exclusively.*

That third element, exclusivity, is powerful. You need to feel that you're the only one who could have perpetrated the act. Once you suspect that someone else might be responsible for something that otherwise you would think was your own, you can easily abandon your sense of will. In fact, we all have a strong tendency to be on the lookout for agents other than ourselves to be responsible for the events and actions we see around us. The most amazing—and funny—demonstration of this goes back more than sixty years.[3]

The experimenters were psychologists Fritz Heider and Mary-Ann Simmel. They painstakingly assembled a two-and-a-half-minute movie (frame by frame) showing the movements, as seen from above, of some geometric figures: two triangles (one large, one small), a small circle and a rectangle with part of one side turned outward, as if you were looking down on a blueprint showing a room with a door ajar. The geometric figures moved in

---

* If you're one of those who believes that ouija boards are a way of communicating with the spirits, you should know that table-turning, a nineteenth-century phenomenon, was believed to be the same thing. Several people would sit at the table, and soon it would begin to move, even rotate rapidly, without anyone believing they had anything to do with it. The table would even answer questions by tapping its leg. However, the great English chemist Michael Faraday used force-measurement devices to show that it was the people's hands, and nothing else, that caused the table to move.

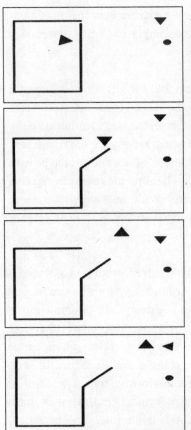

These are frames from Heider and Simmel's 1940s film of geometric figures moving around in a human-like way. At least, that's how it was seen by most of the students who were asked to describe what they saw. It was a good example of how we like to attribute thoughts, feelings and will to anything, whether it's living or not.

and around this "room," sometimes passing each other, sometimes colliding, sometimes appearing to "shut" the "door." Groups of students assembled to view this movie were given different tasks by Heider and Simmel: one group was simply to "write down what happened in the picture"; another was to identify the movements in the film as the "actions of persons" and then go on to answer a series of questions, such as "What kind of person is the little triangle?" or "Why did the circle go into the house?"

The responses were amazing. Of thirty-four subjects in the first group, only one described the movements entirely in geometric terms. Every other student saw the figures as beings eerily similar to humans. Here's a sample description of the movements (and remember, they were asked simply to "write down" what they saw):

A man had planned to meet a girl and the girl comes along with another man. The first man tells the second to go; the second tells the first, and he shakes his head. Then the two men have a fight, and the girl starts to go into the room to get out of the way and hesitates and finally goes in. She apparently does not want to be with the first man. The first man follows her into the room after having left the second in a rather weakened condition leaning on the wall outside the room. The girl gets worried and races from one corner to the other in the far part of the room.

If that strikes you as a reasonable interpretation of a couple of triangles and a circle moving around, apparently you're only human. But remember that it could be described as simple movements of Euclidean figures, as the one student out of thirty-four actually did: "A large solid triangle is shown entering a rectangle. It enters and comes out of this rectangle, and each time the corner and one-half of one of the sides of the rectangle form an opening. The another, smaller triangle and a circle appear on the scene. The circle enters the rectangle while the larger triangle is within. The two move about in circular motion."

Of course, that does sound a little stilted, but she really was the only one who did exactly what she had been asked to do. The second group of students, who were asked to interpret the movements as the "actions of persons," had no trouble doing so (even the first group had no trouble doing that!) and came up with some remarkable answers for the questions that were posed. For instance, in response to "What kind of person is the big triangle?" the following adjectives were chosen: *aggressive, warlike, belligerent, pugnacious, quarrelsome, troublesome, mean, angry, bad-tempered*—the list goes on and on. And remember, we're talking about triangles and a circle moving in and out of a rectangle.

This experiment was published in 1944, long before psychologists were chasing consciousness (at least publicly), and Heider and Simmel were not seeking to explain anything about free will, but they did point out that when the students interpreted the movements of the figures in terms of the acts of persons, they made sense of the film. In their words, the movements "no longer follow each other in an arbitrary and unconnected way. . . . They are meaningfully embedded in our picture of reality." Could there be a more vivid demonstration of our tendency to identify some agent as the cause of any action we see, even if the agent has to be a bellicose triangle?

All this brings us into a weird and slightly hallucinogenic world where events have to be controlled by something, whether that is a spirit, god, invisible companion, tree or rock—just about anything other than ourselves. It also looks forward to Julian Jaynes's proposal from the mid-1970s that in the days before we humans were fully conscious, we *were* automatons, choosing to take action only when bidden to do so by voices generated by the right hemisphere of the brain.* Jaynes argued that these voices were attributed by the people who heard them to the gods, whose images were in turn made more real by the creation of statues. How different is the idea that our actions are undertaken in response to orders delivered by a hallucinated voice from the claim that our actions are undertaken unconsciously and only then revealed to consciousness?

It comes down to this: when you experience some sort of event, your brain's first assumption is that some agent—an alien, a deity, someone else or yourself—caused that event to happen. If it's also true that if just before that event—*just* before—you did something that normally might actually cause an event like that to happen, then your brain will quickly settle on the explanation that this is exactly what happened: you actually *did* cause the event. It needs only that much evidence to make the decision. In

* See Chapter 11.

Heider and Simmel's experiment, people couldn't deal with a movie of geometric shapes moving around: they *had* to imagine that they represented people.

If Wegner is right, and our ironclad feeling that we are the action-eers of the things we do is so fragile that it can be confused or eliminated by simple experiments, or even by paranormal-friendly surroundings, where does that leave our conscious will? Wegner argues that there is a strong possibility that the only role our conscious minds play in free will is to find out what decisions or actions are underway *after* they have been taken. So when you decide to do something, you are actually just becoming aware of an unconscious decision already taken; then when the act is initi-ated, you are made aware, again belatedly, of the unconscious prompt for action. Marvin Minsky put it succinctly in his book *The Society of Mind*: "None of us enjoys the thought that what we do depends on processes that we do not know."[4]

But why would this be? Why would our brains bother to main-tain this powerful illusion of conscious will if we really don't have it? Why not simply go about our business unconsciously, instead of fooling ourselves into thinking we're really in control?

Wegner argues that the illusion of conscious will is maintained because it is important for a social species like ours. Membership in humanity requires some sense of responsibility, and to be responsible you must acknowledge authorship of your actions. But you can't really do that unless you believe that you actually *are* the author.

It is likely no accident that there are many cases of people try-ing to absolve themselves of responsibility for doing something stupid by arguing that their conscious will was temporarily sus-pended: "I dunno what happened—I just snapped"; "I acted on impulse"; "I wasn't myself"; "I was beside myself with rage." Rage is a good one: how many people who have succumbed to road rage would be happy to testify that they carefully and consciously made the decisions to give another driver the finger, swear and then try to cut him off?

By believing that you actually are where the buck stops, you are establishing that you are a participating individual in society. It would be very difficult for others to interact with you, to lean on you or to question you if there were no one home, no one to interact with, no one to depend on.

That would be a sensible reason for the brain to establish the illusion of conscious will, but there is really no evidence that is true. In fact, the idea that we have no conscious will has not yet completely taken hold in the community of consciousness researchers. But once you reflect on it, it is plausible—isn't it?

How Full Is Your Tank of Free Will?

In a strange twist to the whole free-will-versus-determinism debate, Roy Baumeister of Case Western Reserve University and his colleagues have asked the question, just how much free will do you have? They don't come right out and call what they're investigating "free will," but it's the next best thing. They refer to "deliberate, conscious controlled responses by the self" and to "making choices and decisions, taking responsibility, initiating and inhibiting behaviour and making plans of action and carrying out those plans." Sounds like acts of will to me. What makes Baumeister's research intriguing is that his experiments suggest that we do not have a limitless capacity to do these things—that in fact, one act of free will, or volition, makes it more difficult to perform another.[5]

Here's an example. Student participants were seated at a table with containers of chocolate chip cookies and chocolate candies in one and red and white radishes in the other. They had been told they were part of a study on taste perception, and each student was assigned to either the radishes or the chocolates. That meant that each was restricted to eating a small number of one or the other, with no sampling of the other dish allowed. The presumption going into this experiment was that it would be more difficult for those in the "radish" group to restrain themselves from eating chocolates than it would be for the "chocolate" people to resist the radishes. The trick was, how to evaluate that difficulty?

After being seated in the room with the food, the students were encouraged to sample the food they were assigned to, then after about five minutes an experimenter came back, gave them a couple of questionnaires, then told them they would have to wait about fifteen minutes for the sensory effects of the food to wear off. During that time, the students were asked to solve some puzzles (they were told this was part of another experiment). But the puzzles were actually impossible—they couldn't be solved. For example, in one puzzle the students had to trace a geometric figure without lifting the pencil from the paper, but the lines were drawn in a way that it couldn't be done.

But the time each student spent on the puzzle before abandoning it in frustration was the key to the experiment. There was a significant difference in persistence between the two food groups, and, as you might expect, the "radish" group, apparently mentally worn out by the demands of resisting the chocolates, gave up on the puzzles significantly sooner than the "chocolate" group. (Amazingly, not one of the radish group actually cheated and ate any chocolate, although one went so far as to pick up a chocolate chip cookie and smell it, only to put it back—you can almost feel the will power draining away.)*

Partly to forestall criticism that persistence in trying to solve an impossible puzzle is actually kind of stupid and that giving up early makes sense, Baumeister's team used a slightly different test where the puzzle—trying to solve anagrams—was actually possible. In this case, students watched movies and were asked either to suppress any emotion they felt or to let their emotions run free. Again, those who had had to work hard to suppress emotions just didn't have the mental wherewithal to persist at the anagram task, whereas people who had experienced no such stress stayed at them longer.

Fascinating experiments, and ones that make sense when applied to the realm of consciousness. If acts of will draw on some finite resource—if you only have so much free will in the tank—it makes perfect sense to limit the number of occasions when such stressful thinking is required. If

---

* To eliminate the possibility that the chocolate consumers were just riding a sugar high that allowed them to persevere with the puzzles, Baumeister included a control group who ate nothing. They persisted just as long as the chocolate people.

conscious decisions were required every minute of the day and the "ego resource," as Roy Baumeister calls it, were depleted so easily, we would be at a virtual standstill by lunchtime. From this point of view, we should be grateful that the unconscious does so much for us: we just aren't capable of having it any other way.

CHAPTER TEN

Animal Consciousness

I N 1974, Thomas Nagel of New York University published an essay that would become one of the most quoted in the pursuit of consciousness. It was called "What Is It Like to Be a Bat?"[1] The curious thing was that Nagel is not a chiropterologist, a zoologist or even a biologist; he is a philosopher. He wasn't writing about what it would be like to inhabit the brain of an animal whose picture of the landscape and the living things in it is built up not from vision, but from echolocation; rather, he was arguing that it would forever be impossible to do so. Nagel was bolstering the philosophical argument called the "other minds" problem, which argues that because consciousness is subjective—I can't know exactly what it feels like to be you, and vice versa—it likely will never be understood no matter how much scientific data are collected, no matter how sophisticated the sensors and computers applied to it. We'll never be able to make objective what is uniquely subjective.

Nagel chose bats for his purposes rather than insects or fish because he recognized that if he chose an animal that was too dissimilar from humans, whose nervous system was too much smaller or constructed from radically different blueprints, there'd be questions about whether that animal was conscious at all (in which case there would be *nothing* that it would be like to be one). Even though bats' predominant sensory system looks pretty weird to us, they *are* mammals and, aside from echolocation, not too different from us. Nor was he asking his readers to pretend they had webbed wings, ate insects on the fly and hung upside down in attics. As Nagel pointed out, that would be asking us to imagine what it would be like for us to be bats, but he was really after was what it is like for a *bat* to be a bat—a very different question. Even if, Nagel suggested, one could be gradually transformed, even neuron by neuron, into a bat, it would still be impossible to predict what the final product would be like.*

In the end, Nagel's bat was a device to address the bigger question of whether or not it is even possible to examine consciousness scientifically. He argued that this would be unlike any other science, where reducing observations and data to the basics brings you closer to objectivity—in this case, you can't get a more objective point of view of the thing, which is essentially subjective. He asked, "Does it make sense, in other words, to ask what my experiences are *really* like, as opposed to how they feel to me?"

But while Nagel assumed (to make his point) that bats are conscious, you could take his title the other way, and wonder if indeed there is something that it's like to be a bat. We may assume (but we don't know) that they actually are conscious. If we are going to make sense of human consciousness, it would be very helpful to know just how widespread consciousness is in the animal kingdom.

Is it reasonable to assume that mammals are conscious? Birds? Reptiles? Fish? I'm sure you have a different and instantaneous reaction to each of those examples, and if you're typical, you would probably grow increasingly doubtful as you worked your way through that list. Mammals, probably; fish, I don't know. Insects,

---

* I suspect that even if you could turn into a bat then back into a human, you'd be unable to describe what temporary bathood was like, not only because there literally wouldn't be words to describe it, but because you would no longer have the right kind of brain.

probably not. But if you were to dig a little deeper and examine your reasons for making such judgments, you'd have to admit they aren't really well founded but are more than likely based on the fact that each of those groups exhibits behaviours that diverge more and more from ours (you might also make an argument based on the size of their brains, ranging as they do from larger to smaller).*

But you need to do better than that, because if I had an opposing opinion, I could bring just as much data to support it as you could (that is, not very much). For example, you might argue that your cat Fluffy enjoys meaty chicken chunks better than ground hamburger, because she eats it faster. "Enjoys"? Does that mean that little Fluffy is reflecting on the deliciousness of her food, or is it just that she scarfs down the chicken faster because it's easier to swallow, or that she had more of it when she was a kitten, or that there is some chemical combination in the food that, quite unbeknownst to her, triggers faster eating, or is it just that you haven't been watching closely enough?

What is the single most difficult problem you would have establishing that there is something that it is like to be Fluffy? That's easy: she can't tell you. So how can we say she's conscious? We can have the faith that she is, but that's not science. (See Gary Larson's famous cartoon of the cat listening attentively to its owner and registering the loving words as "blah blah blah blah blah.") Animals cannot discuss their experiences with us, so it's much harder to tell if they're conscious. It's true that even if they could, we still might not be absolutely sure, but the same qualification applies to ourselves—all we have to go by is our testimony that we are conscious—and that's usually good enough.

---

* How much does size matter? Dolphins and whales are popular candidates for smarts, at least partly because they have huge brains. But dolphins' brains are structured differently from ours, and actually have fewer neurons than you might have expected; one estimate suggests they have about as many as chimps. That's not bad, but it doesn't support the idea that they're super-intelligent. If you're going on brain size, why haven't elephants been accorded the same sort of respect? They have huge brains, structured like ours, and even if you factor in the differences between an elephant brain and a primate brain, they should have the same number of neurons as we do.

Animals could do very tricky things that might simply be assemblies of behaviour that they really don't think about but perform automatically. Just as I was writing this chapter there was a fascinating report from England about the crows of the South Pacific islands of New Caledonia. These birds are famous for using twigs to get at food hidden in crevices. Two crow chicks were taught how to do this in the absence of their adults, and two were left to their own resources. Amazingly, all four picked up the tool-using habit with no problem, and the ones that learned it on their own were just as skilled as the ones that had been taught.

The fact that this skill apparently has a sizeable genetic component doesn't preclude the birds' thinking about it, but it does at least open the door to the possibility that it could be an elaborate, preprogrammed behaviour, executable with no need of reflection on the part of the bird. I'm not saying that it is that way, just making the point that there is that possibility: if you're going to establish beyond a reasonable doubt that animals are conscious, such possibilities of unthinking behaviour have to be eliminated.

Unfortunately, many seemingly purposeful animal behaviours can be cast into doubt this way. Take the famous example of a nesting bird that, at the close approach of a potential predator, begins to flutter around on the ground as if its wing is broken, luring the intruder farther and farther away from the nestlings. It *looks* deliberate. In fact, however, Carolyn Ristau at Barnard College has shown that the behaviour can be tweaked depending on the circumstances.[2] Piping plovers nest in the open on sandy beaches, and she observed that when a human simply walked by at a distance, apparently without noticing the nest, the plover remained still. But if the same person paused and looked, the bird immediately began its broken-wing routine. Eventually the plovers became familiar with individual humans who had previously been judged to be safe, and remained on the nest even when they would normally have been provoked.

These sound like thinking behaviours, but let's face it: we are prone—if not programmed—to interpret the behaviour of other

animals in our own terms. I mean if we can give triangles person-
alities, as in Heider and Simmel's experiment in the previous
chapter, we can certainly do the same for lions and blue jays. In
addition to that bias we bring, it is conceivable that the plovers
are acting out an evolutionarily tuned decision tree, in which a
certain action by an approaching animal triggers one response,
the next a second and so on. Such a series of actions would
require a long and involved evolutionary history, but every step
should make sense, beginning with the first one: birds that didn't
bother to leave their nest when predators approached didn't live
to reproduce (nor did any of their eggs survive); birds that left the
nest *and* lured predators away succeeded on both counts.

That kind of skeptical thinking about animal consciousness just
naturally fell out of behaviourism, which dominated psychology for
decades in the middle of the twentieth century and forbade speculation about the inner goings-on of the mind. The behaviourist
approach to animal consciousness began with a champion. B.F.
Skinner might be behaviourism's man, but Conwy Lloyd Morgan
set the tone before him. In 1894, in the book *An Introduction to Comparative Psychology*, Morgan set out Morgan's Canon: "In no case
may we interpret an action as the outcome of the exercise of a
higher psychical faculty, if it can be interpreted as the outcome of
the exercise of one which stands lower in the psychological scale."

The fact that this statement has been elevated to the status of a
canon shows just how seriously it was taken. Maybe too seriously:
all Morgan was really asking was that people do experiments rather
than read too much into animal behaviour without enough evidence to justify doing so. In his view, the challenging questions: "will
have to be settled, if [they] can be settled at all, not by any number of
anecdotes—interesting, and to some extent valuable, as such anecdotes are—but by carefully conducted experimental observations,
carried out as far as possible under nicely controlled conditions."[3]

Whether Morgan was read closely and carefully or not, the
impression you get from psychologists looking back is that his
canon put a damper on thinking about animal intelligence and

consciousness. A bold scientist occasionally broke with the rest to argue that consciousness was rampant among animals, especially mammals, but most shied away from (or just didn't agree with) making that argument. However, the domination of the Lloyd Morgan canon was weakened by the 1960s, and over the last ten or fifteen years, scientists have become much more willing to speculate about how widespread animal consciousness might be. Even so, the claims of consciousness have been largely focused on the predictable: the great apes, dolphins and parrots.

It helps us be receptive to the idea of great ape consciousness that they have brains like ours, although slightly undersized. But the hard evidence for consciousness in their case is the performance in intelligence and language tasks by Kanzi the bonobo, Ai the chimpanzee and Koko the gorilla (although in this last case virtually nothing has been published in the scientific literature, raising a flag for me).

Dolphins too are demonstrably smart, with something close to linguistic capability, putting them in the club as well.* Again, we're okay with that because they have supersized cerebral cortexes (just like us, or rather, a little bigger than ours).

Alex the grey parrot is a different case: he does amazing things, but because it's not easy to line a bird brain up beside a human brain and immediately see that the two are closely related, we're just not sure that what Alex is doing is consciousness the way we understand it. His trainer, Irene Pepperberg, calls it "perceptual consciousness," which implies that Alex is aware of the sensory information being handled by his brain, even though he isn't necessarily aware that he *is* aware. He's acting on his awareness, not reflecting on it.

---

* Sorry for the awkward substitution of "linguistic ability" for "language." You want to write, "They can talk!" but they can't literally talk—they use symbols, and even then, most linguists would want substantially more than they've seen before they'd grant that an ape was using "language." They'd want the complicated clause-driven speech of humans, complete with recognition abilities adept enough to realize that the two sentences "Bill gave Mary the cup" and "Mary was given the cup by Bill" mean the same thing. So I waffle by writing "linguistic abilities," meaning to me that at least some of the key elements of human language are there.

Alex's oddest responses in experiments are his most interesting. In one test, the psych-lab version of the old shell game, Alex was shown a prized cashew, which he then watched as it was being hidden. But the experimenters surreptitiously substituted a food pellet for the cashew, and when Alex chose what should have been the right container and found the much less favoured food pellet, he responded by narrowing his eyes at the experimenters, a reaction they knew from experience to be anger. On a second trial, he banged his beak on the table, apparently in frustration. The experiment did test his memory and his awareness that objects, even when they disappear, may still be present, but the results also revealed not only the capacity for surprise, but an emotional reaction to that surprise.

Another experiment demonstrated that even Irene Pepperberg, who has worked with Alex for years, can't always read his mind.* In one experiment he was given seven different items of various colours, shapes and materials and asked to label the colour of the one that was, for instance, wooden and square. He gave the six possible wrong answers in a row, then proceeded to give the same six wrong answers again, one by one. If he were just choosing answers randomly, he should have stumbled on the correct one, but for some weird reason he was being perverse. As Pepperberg has said, "The bird must have a representation of the labels of all attributes, integrate that information with a search for the appropriate item (which requires combining representations of two attributes), encode the correct attribute label, then specifically avoid uttering that label to produce all other relevant labels for 12 trials. We suggest some awareness drives this behavior."[4]

Pepperberg wondered if deliberately giving wrong answers might be one way a parrot can capture its caretaker's attention in an unexpected (at least for the caretaker) way. Answering correctly

---

* I know from first-hand experience Alex is no automaton. When I visited Pepperberg's lab to shoot some video of Alex doing his thing, he proceeded not to do his thing, but in a way that made you suspect that this was his decision, not his inability.

would be logical; if Alex knew that but chose to answer illogically, he was demonstrating some pretty powerful reasoning—possibly, but unfortunately not provably, conscious. In a way, Alex represents the fringe: almost no animals further away evolutionarily from humans have ever seriously qualified for consciousness.

So while these species, or at least individual representatives of them, have convinced most experts that there is consciousness there, there is still the issue of how many others might be conscious as well and, more important, of how to establish that fact objectively—how to get from animals actual data that point to consciousness. Some scientists are meeting that challenge by climbing back down the evolutionary tree and arguing for a serious program of consciousness detection in the animals you find there.

Anil Seth, David Edelman and Bernard Baars of the Neurosciences Institute in San Diego, California, think that in addition to attempts at communication like those already described, there might be two other approaches worth trying.[5] They are (1) deciding what parts of the human brain (and the patterns of activity they exhibit) are involved in generating consciousness, then checking to see if other animals' brains have the same structures and patterns and (2) testing the responses of animals in certain situations to see if they are comparable to the conscious responses of humans in parallel circumstances. Both of these have yielded some suggestive evidence, but neither has yet nailed down the existence of animal consciousness unambiguously.

The first approach, that of comparing brains, has a significant problem before it even gets off the ground.* So far there has been no one structure identified in the human brain as the key

---

* One of the most outspoken champions of animal consciousness, the late Donald Griffin, actually reversed the brain argument by suggesting that maybe it was only big brains that needed to process so much information unconsciously, and that small brains, lacking such an informational burden, might actually be *more* reliant on consciousness. I suspect few neuroscientists will take that idea seriously; almost everyone suspects that consciousness emerges only when there is a critical mass of brain stuff present.

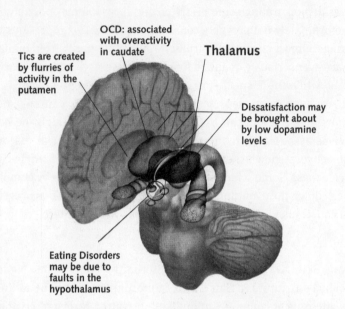

OCD: associated
with overactivity
in caudate

Tics are created
by flurries of
activity in the
putamen

Thalamus

Dissatisfaction may
be brought about
by low dopamine
levels

Eating Disorders
may be due to
faults in the
hypothalamus

While it is unlikely that any one part of the brain will turn out to be the "hub" of consciousness, the thalamus would come close. Myriad circuits of neurons are routed through it, and it controls not only the flow of neural traffic up to the all-powerful cerebral cortex, but also its activity.

to consciousness. In fact, much of the modern scientific program is the search for what are called the "neural correlates of consciousness," those groups or circuits of neurons that are responsible for generating conscious experience. That said, there is evidence that, at least in humans, complex circuits running between the thalamus and the cerebral cortex are at least necessary for consciousness, if not sufficient.

The thalamus is the shape of a quail's egg, and about the size of two walnuts held together. It is a mass of neurons, millions of them, buried in the middle of the brain just under the cerebral hemispheres. Not only is it the place where most sensory information makes a final stop before moving upward to the cerebral cortex, but it also transmits signals from the cortex to the muscles *and* controls the flow and intensity of information to and from the cortex. The thalamus has been called the brain's switchboard, a metaphor that dramatically understates its role.

Cutting the fibres of one small part of the thalamus, the cluster of neuron groupings called the intralaminar nuclei, causes coma.* There are very few brain areas that, when damaged, have such a dramatic and direct effect on consciousness. They not only participate in the thalamic control of neuronal traffic flow, but also appear to tweak the activity level of the cerebral hemispheres. They play a generalized role: they're not responsible for the traffic of specific sensations or ideas, but you can't be aware without them. If you are startled by the sudden appearance of a threat, all your senses are heightened, your brain on alert, thanks in part to the intralaminar nuclei.

But the thalamus isn't all there is to it. Some scientists argue that recordings of electrical activity from the human brain can be used to differentiate conscious brain activity from its unconscious counterpart, and that if such electrical patterns were recorded from animal brains as well, it would at least suggest the presence of consciousness. For instance, conscious brains are characterized by bursts of simultaneous activity in widespread areas of the brain, often linked to incoming sensory information. Such patterns are easily distinguished from unconscious brain states like that of slow-wave sleep, which, as the name implies, is a steady, slow and repetitive electrical pattern, largely unresponsive to the senses and producing no sudden neuronal flare-ups.

---

* You might remember the case of Karen Ann Quinlan, who at the age of twenty-one in 1975 overdosed on alcohol and Valium and lost consciousness permanently. Her family's fight for the right to withdraw life support became headline news in the United States. They won that fight, but when the plug was pulled Karen survived, and lived for another ten years in a nursing home. The autopsy revealed that little if any harm had been done to her cerebral hemispheres, but there was extensive damage to the thalamus, including the intralaminar nuclei. The more recent Terri Schiavo case was different. Brain imaging and the autopsy showed that her cerebral hemispheres had been largely destroyed.

Seth, Edelman and Baars are saying, look for brains that have dense connections between the thalamus and the cortex (or the structures most like them) or brains that exhibit electrical activity typical of awareness. There you might find consciousness. The best part of this approach is that it encourages casting the consciousness net very wide. Finally we can talk about some animals other than the great apes! Seth, Edelman and Baars point out that the fundamental circuitry running from thalamus to cortex—circuitry that is necessary (but perhaps not sufficient) for consciousness—has been in place since mammals first appeared in the evolutionary record, more than 100 million years ago. That is a startling thought: there have been very few suggestions that animals in the age of dinosaurs were thinking, aware, conscious creatures. Just as boldly, Seth and company advocate taking a close look at birds and cephalopods, specifically octopuses.

Birds have, of course, been nominated for consciousness before. I described Alex the grey parrot, but you can find plenty of additional research showing that birds can remember where they've hidden dozens or even hundreds of seeds or nuts, and can even remember which hiding place contains the more easily perishable food, so that it can be collected first. Ravens are so far the only bird species smart enough, when faced with a piece of meat dangling on a string below a perch, to haul up short lengths of string with one claw and pin them down to the perch with the other, eventually getting to the meat.

But there are less well known examples that to me are even more intriguing. For instance, in 2001 researchers at Cambridge University published the results of their experiments with thieving scrub jays, birds that live in the southwestern United States and in winter feed largely on pine nuts, seeds and acorns.[6] They also steal from the seed caches of other birds, such as Clark's nutcrackers. And they are perfectly willing to steal from other scrub jays as well. What Nicola Clayton and Nathan Emery were able to show was that these jays seem highly aware of their social situation and are capable of something close to deceit.

If you give a scrub jay a bowl of wax worms and a sand-filled tray, the bird will hide the worms in various places under the sand—no surprise there. But if a second scrub jay is watching while this happens and the jay doing the caching has its own (guilty) history of pilfering other birds' caches, it will stand around and wait for the watcher to leave, then move the worms to a different hiding place.

Now just take a moment to reflect on what this says. A scrub jay waits for an onlooker jay to leave, then re-hides all its worms. Why? It is conceivable that this bird's brain automatically and unthinkingly works its way through a program of calculations and actions that end up by prompting the bird to change the location of the worms. If that were true, the bird would definitely remember that the worms were in new locations but would have no idea why it moved them. Let's face it: it seems much more reasonable to suppose that the bird doing the caching realizes that the jay that's watching is capable of doing in the future what it has itself done in the past: stealing the worms. So it moves them.

Emery and Clayton called it "the first experimental demonstration that a non-human animal can remember the social context of specific past events, and adjust their [sic] behaviour to avoid potentially detrimental consequences in the future," but it seems more dramatic to me than that. These birds seem to be doing exactly what's necessary to prove they have "theory of mind," the skill of putting yourself in someone else's place to figure out what they're going to do.

"Theory of Mind"

In the last ten years there has been a lot of attention focused on the psychological term theory of mind. It refers to the understanding we have that other individuals have their own minds, and that while most of the time those minds are concentrating on things that are different from what's in your mind, under the right circumstances you can "know" what they're thinking about. If you can't take your eyes off the chocolate cake in the bakery, I'll take note of that. Maybe I could surprise you pleasantly by sneaking back and buying the cake. Maybe I could surprise you unpleasantly by pretending that's what I've done then revealing the cruel truth at the last minute. Regardless, it's "theory of mind" that lets me into your mind to know what you're thinking.

Some animal experts are convinced that theory of mind is closely related to a sense of "self" and that, in turn, having a sense of "self" is part of consciousness. So any evidence that an animal can empathize with another, deceive it or in some way "know" what's going on in another's mind shows that that animal has theory of mind and so can be considered to be one step along the way to consciousness.

The study of theory of mind in animals began with clever experiments by Gordon Gallup Jr. nearly forty years ago. Gallup wanted to find out exactly what chimpanzees were looking at when they gazed at themselves in a mirror. Was it just that there was a chimp face in the mirror, or was it that it was their own face? Gallup used chimps that were very familiar with mirrors; he anaesthetized them briefly and, while they were out, put a coloured mark on their forehead. When the chimps awoke and looked in a mirror, the ones who had been marked touched their own forehead repeatedly. Chimps that had not been marked did not; chimps that had been marked but were not given mirrors didn't either. Orangutans and dolphins, but not gorillas or monkeys, have also passed the mirror-recognition test.

Gallup has elaborated on this in the years since, and he puts forward a two-step argument: animals that can recognize themselves in a mirror also possess self-awareness, and self-awareness leads to theory of mind. You have to know who you are (and what you're thinking) first before you can tell the difference between what's in my mind and what's in yours. Gallup argues that self-recognition as embodied by the mirror test is necessary for self-identity, the awareness that you are you, and that self-identity in turn leads to theory of mind. You can't feel empathy or sympathy, you can't deceive, unless you know what's going on in the other's mind.

Theory of mind is on the road to consciousness, although they are not the same thing. But it's difficult to imagine how some sort of elaborate but unconscious set of brain circuits could automatically and unthinkingly interact with other members of the species in all the right ways without some awareness of what was happening. At the very least, understanding what you might be thinking and therefore what you might be doing next requires that I be able to think about the future, something that some experts identify as an important feature of consciousness.

Gallup and former student Julian Paul Keenan have gone further and identified the right hemisphere of the brain as the place where self-recognition and theory of mind are located. They asked volunteers both to view a series of faces (their own alternating with a famous face) and to judge the feelings of people from still pictures of their faces. The self-face test turned on a lot of different brain areas, two of which were also activated during the "feelings" test. Gallup and Keenan now argue that these areas—the middle and superior frontal gyri in the right hemispheres, for you anatomy buffs—are the seat of self-recognition and theory of mind. Because these are areas that have evolved relatively recently (big frontal lobes are pretty much a great-ape-or-later thing) you might think that these studies suggest than only animals with a large and elaborately folded cerebral cortex could be self-aware. But don't forget those scrub jays.

This research isn't exactly making overnight converts. One of the most contentious issues is the interpretation of the mirror-recognition test. If an animal can recognize itself in a mirror, does that mean it has a sense of self-awareness? Gallup thinks so, but others argue that it's possible just to do the brute-force face-recognition processing without anything else. No need for "self-awareness of the introspective type."[7] An animal mind of this type might have no idea what is on its mind, or whether it even has one, but it would be able to recognize its own face. How this ability would develop in the absence of mirrors is somewhat mysterious.

Also by putting consciousness in the right hemisphere Gallup and Keenan are leaving language (usually found in the left hemisphere) out of the picture. This is going to require a lot of convincing, because most experts are confident language plays an important role in human consciousness, the most obvious and easily accessed example being the nonstop interior monologue called "inner speech."

Finally, split-brain patients (see Chapter 13) are very difficult for the right-hemisphere theory because these patients' isolated left hemispheres are fully conscious, and are in fact easier to communicate with (because they contain language) than the right. But if self-awareness leading to consciousness is located in the right hemisphere as Keenan and Gallup claim, how does it make its way over to the left? As far as I know, there are no theories on the go, although in the case of these patients, surgery was only performed relatively late in life, long after a sense of self would have been established in both hemispheres.

Octopuses are a different matter: they are much more intelligent than you would expect from an invertebrate, but are they conscious? Certainly it's hard to find any parallels between the octopus brain and our own. In fact, there are more neurons in the octopus's eight tentacles than there are in its brain. Nonetheless, that brain shows similar kinds of electrical activity to the human brain, although for consciousness' purposes the octopus brain is so different from ours that it's hard to know exactly where such recordings should be taken.

Octopuses can negotiate mazes skilfully, sometimes appearing to the scientists working with them as if they consider the layout of the maze before attempting it. They can learn to take the top off a glass jar to get at the prey item—a crab—inside. But it is the way they succeed at this task that makes some reluctant to declare them conscious. Even after many repetitions, octopuses take the same approach to the jar, pouncing on it as they always do when attacking prey. In that sense they seem not as flexible, or reflective, as grey parrots. Communicating with them seems to pose a much bigger challenge than with the other animals studied so far, although they do have an amazing ability to change colours and patterns of colour on their skins, and who knows? They might be able to use that ability to tell an experimenter they're angry just as Alex did.

The other suggestion by Seth, Edelman and Baars is to see if animals react similarly to humans in parallel situations; if humans react consciously, it could be argued that the animals are as well. There are two experiments in particular that demonstrate how this would work. One, ironically, was an experiment done on humans, by Robert Clark and Larry Squire.[8] It was totally straightforward, just a replay of the old Pavlovian conditioning studies. They played a tone, then followed it up with a puff of air aimed at the eye, forcing the subject to blink. Of course, after many trials you'd blink just at the sound of the tone. Clark and Squire found, however, that varying the time lag between tone and puff revealed something curious. In one version the tone stayed on until the air puff; in the other it ended anywhere from half a second to a full second early. During the whole thing the subjects watched a silent movie.

The results were intriguing. Everyone responded to the initial part of the experiment, in which the tone continued until the puff; everyone acquired the Pavlovian eye-blink in response to the tone. But in the second part, in which there was a gap between tone and puff, only those people who were aware of the sequence of events—that is, who could testify that the tone sounded before the puff—became conditioned. People who apparently had no idea of that time sequence never became conditioned to blink after hearing the tone. Awareness was a key component; without it, there was no conditioning. Further elaborations of this experiment showed that amnesiacs who had suffered damage to the hippocampus, an area of the brain crucial for laying down memories, were incapable of the delayed conditioning: They simply forgot what happened. Also, people who had had to concentrate on a different task at the same time didn't condition either. Both cases support the idea that in the absence of awareness of the timing of tone and puff, conditioning won't occur.

What is significant about this finding is that rabbits appear to be in the same boat: do the same experiment with them, and those with damaged hippocampuses fail to become conditioned, but normal rabbits do. Draw a parallel between the two animals, and here's what you get: an animal has to be aware of the sequence of events to have conditioning work—rabbits and humans alike. Are rabbits aware? Or are rabbits *conscious*?

Christof Koch, who together with the late Francis Crick kicked off the search for the neural correlates of consciousness about fifteen years ago, has asked whether, given that even fruit flies can be conditioned in this way, we should consider the possibility that *they* are conscious. He points out that we have no idea how many brain cells it takes to be conscious, a billion, a million or ten thousand.

Could Insects Be Conscious?

Christof Koch suggested that fruit flies might be conscious because they can be conditioned in a way that seems to require awareness of what's going on. But—fruit flies? Besides the sheer unlikelihood of it, there are some reasons to believe that insects are not conscious. They have crazy nervous systems. There's nothing in them that shouts "consciousness," but we have to admit at the same time that we don't really know what we're looking for. Also, they seem to ignore crippling injuries, like the loss of a couple of legs. That is a very robotic reaction. And it's also true that some insects' adult life is only a few hours, or even minutes, hardly enough time to bother being conscious.

All that's true, but is it possible that because insects are so alien to us we're discounting their chances? One of my favourite arguments in this regard was written by philosopher Daisie Radner in her book *Animal Consciousness*. Radner begins by retelling the story of the yellow-winged wasp, *Sphex flavipennis (ichneumoneus)*. This wasp was the central character in a study by legendary entomologist Jean Henri Fabre in the nineteenth century.[9]

The female wasp lays her eggs at the end of a tunnel in the ground, complete with a living cricket paralyzed by the wasp's sting. When the larva hatches, it can eat the cricket at its leisure. The adult wasp goes through a sequence of actions when she arrives at her nest with cricket in tow. She leaves the cricket at the mouth of the burrow, disappears inside to check out the nest, then returns and drags the cricket in with her. Then she lays her eggs on the cricket.

Fabre intervened in this sequence by waiting until the wasp went down the tunnel on her inspection tour, then moving the cricket a few centimetres away. When the wasp emerged, she would cast around, find the cricket, move it back to the mouth of the tunnel and then disappear again down the tunnel to check out what she had just checked out a few seconds before. Forty times in a row Fabre moved the cricket, and forty times the wasp engaged in the same loop of activity. The same *mindless* loop of activity? That is what most commentators think. That is certainly what Fabre thought, remarking that "instinct knows everything, in the undeviating paths marked out for it; it knows nothing outside those paths." It has even entered the philosophical literature as the term "sphexishness."

Daisie Radner took issue with this straightforward interpretation of Fabre's experiment with *Sphex*. She quotes Fabre himself as asking, "Could it not be that, before descending with a cumbrous burden, the Sphex thinks it wise to take a look at the bottom of her dwelling, so as to make sure all is well and, if necessary, to drive out some brazen parasite who may have slipped in during her entrance?" That, of course, is a reasonable explanation of the wasp's behaviour, but it doesn't excuse the endless repetitions. Or does it?

Radner offers a human analogy in the wasp's defence: "One can easily imagine a case in which it would be wise to repeat forty times or more the process of looking before you leap. You are on the fifteenth floor of a burning building. A safety net is being held for you below. Forty times the smoke and falling debris drive you back. Each time you approach the ledge you check to make sure that the net is still there, for the firemen could have moved it to rescue someone else in the meantime."

Radner was not arguing that *Sphex* is conscious, but she was questioning the knee-jerk assumption that the wasp is, well, sphexish. The wasp might just be unbelievably thorough, or a touch paranoid, or incredibly forgetful. (Would an alien investigator dismiss the likelihood that we're conscious because we leave the house, then enter it again to shut off the stove that we just shut off? Would the same investigator draw the same conclusions about obsessive-compulsive behaviour?) Can't wait for those experiments with the fruit fly.

In addition to conditioning experiments, brain imaging holds some promise in being able to establish who's conscious and who's not. It has proven to be relatively easy to show that monkeys experience the same kind of visual consciousness as we do. These experiments involve something called binocular rivalry. A monkey sees one thing—say, a face—with one eye, and something completely different, like a sunburst pattern, with the other. When humans view different things with each eye, they are conscious only of one at a time, in the same way that only one orientation of the Necker cube is seen at any time. The timing of the switch from one image to the other is out of our control—it just happens.

Monkeys can be trained to indicate which of two images they're aware of at any moment by pushing one lever or another. It is clear from these experiments that monkeys experience first one image then the other, at about the same pace as we do. The only reasonable explanation is that they're conscious of the images as we are.

So there is more than just our intuition to go on. There are experiments that point to animals being aware of the world around them. But this is exactly where the story starts to blur. You likely already have the impression that consciousness may not be just one thing, at least when it comes to animals. I keep having to refer to "human" consciousness because most scientists who take the idea of animal consciousness seriously are ready to accept that there are different levels or kinds of consciousness. You'll remember that Irene Pepperberg referred to Alex the parrot as having "perceptual consciousness." That is a straightforward awareness of incoming sensory information. But consciousness could be more complicated, like the difference between feeling thirsty and reflecting on the fact that you are thirsty. Animals might have the first without the second.

Do animals know that they're conscious? Can they say to themselves, "How about that? I'm thinking about that particular kind of dog food that I ate last week"? If they can, they are experiencing an even higher level of consciousness that includes self-awareness and the ability to project into the past and future. Unfortunately each researcher tends to use his or her own labels for these hypothetical different consciousnesses, and the numbers grow. One article I recently read defined eleven different kinds of consciousness.

There is no point listing them all: the general idea is that there's a progression from some sort of dim awareness, the first consciousness to appear in animals (and one that might still exist relatively unchanged among reptiles and amphibians—fish, snakes, turtles and their kin), through more elaborate consciousnesses like Alex the parrot's, finally to the one we know best: ours. Changes in the way these brains think would be accompanied along the way by changes in brain size and organization. It may not be as important (or even possible) to label each one precisely as to acknowledge that animal consciousness might come in forms that are alien to us.

It's also worth noting that while the vast majority of publications on this topic report experiments that narrow the gap between humans and animals, I don't think any of those researchers would deny that there *is* a gap, and it is substantial. What could account for it? It is true that we have proportionally bigger brains, at least in the area that we think is the one that counts: the cerebral cortex, especially at the front. But it's clear from the binocular rivalry experiments that apes and monkeys have similar sorts of awareness to ours. Why then is "culture" for them a question of exactly how to squish a bug, or which tool to use to fish termites out of a mound? If we could answer this question, we would be a lot closer to understanding our own consciousness.

Hoping to learn about ourselves is, of course, the main reason for studying consciousness in animals, but a more important reason has to do with our relationship with them.

It's always easier to kill living things—or mistreat them—if you're pretty confident that they aren't aware of what's happening to them or that they are insensitive to it. But as soon as you grant them consciousness, things get a lot more difficult, because even the most primordial forms of consciousness should include the feeling of pain. Figuring out where to draw the line of consciousness will be difficult to do, but the results might be disturbing.

At the same, time working with animals apparently inspires scientists to think freely about consciousness. A perfect example is the writing of Sue Savage-Rumbaugh, who has worked for years with bonobos, the "second chimp." The bonobo Kanzi has demonstrated an incredible ability to communicate with symbols, and can even understand spoken English. Kanzi *has* to be conscious, and there's no evidence that he is any sort of rare specimen.

Savage-Rumbaugh has used her experience with Kanzi and other bonobos to come to a dramatic conclusion. She is convinced that culture and environment play a huge role in consciousness and that any analysis of experiments with apes has to take that into account. She and her colleagues also believe that consciousness is a fundamental property of the universe, like electrical charge: "Consciousness is a property which the brain manipulates in ways we might conceive of as bending, folding, focusing or magnifying. . . . We suggest that reality is a construction of consciousness molded by forces of the brain shaped by culture."[10] She even argues that culture drives speciation: "A nonhuman primate like Kanzi, reared in a *Pan/Homo* culture, capable of understanding spoken English and uttering lexical English counterparts, should possess a brain which is morphologically different than a brain of a feral bonobo or a bonobo reared without human language, culture and tools."

This is a fascinating idea: how far could you drive the size, shapes and capabilities of the chimpanzee brain with the right environment? Would there still be differences that can't be eliminated? It makes you wish there were another primate on earth, another great ape, but one with a brain and capabilities halfway between modern chimps and humans. What would the inner life be of an animal like that? Still below some crucial threshold for humanness, or halfway there, suggesting consciousness is a gradual thing? While there unfortunately isn't such an ape, we don't yet know how the dolphins and great whales might play a similar role in contributing to our growing understanding of consciousness. From the point of view of brain size and structure, animals like sperm whales might turn out to be the most intriguing, if we could ever figure out how to communicate with them.

When you consider that species like sperm whales are still alien to us, you realize that we've barely scratched the surface of the study of animal consciousness. I'd bet that as our understanding of other species increases, the study of animal consciousness will make hugely important contributions to our understanding of our own consciousness.

A Bat's World

Acknowledging that Thomas Nagel was not actually trying to imagine what a bat experiences in its nightly life (despite the title of his paper, "What It's Like to Be a Bat"), this famous philosophical essay does make you wonder what indeed it would be like. Brock Fenton of the University of Western Ontario is a world expert on bats. He understands their sensory perceptions as few others do, and here's how, according to him, a hoary bat—the most common North American species—experiences its world while in flight:

As I cruise hunting for prey, a nice juicy moth for example, I produce about four pulses of sound (echolocation call) each second. The sounds come from my voice box, and echoes of them rebound from objects in my path—trees, rock faces, and, if I face downwards, the ground. Each time I produce an echolocation call I register details of it in my brain and then compare what I said with the echoes that rebound.

I use information about timing to estimate distance to my moth target, and by comparing information from successive calls, track my progress relative to the flight path of the moth. Because my ears do not allow me to broadcast and receive signals at the same time, I use progressively shorter calls as I close with a target.

Information in the frequency components of the sound and its echoes (the call structure) give me specific details about the target, its shape, the orientation of its wings and its body surface. Beetles, which lack scales,

The hoary bat may be very common, but it's impossible to imagine what it's like to be one.

return a crisp echo, while moths which are covered with scales (or caddis flies with hairs), give more fuzzy echoes.

To adjust my picture of the insects before me, I can alter the timing of my signals. Longer signals allow me to extend my range of operation (perhaps to about 20 m), while shorter signals deliver details about close targets. In longer signals I usually concentrate most energy in the lower frequency components of my signals (around 20 kHz). The wavelengths of the signals determine the details of the targets that I can obtain by comparing pulses and echoes. For longer signals the lower frequency signals give me greater range but less detail. The shorter signals have a mixture of higher (35 kHz) and lower (20 kHz) components, giving me more details about the overall targets.

Using my longer, lower frequency signals, I am pretty much deaf to small insects (like mosquitoes) which just do not come up on my echolocation screen.

As a hoary bat I can see and hear the world just like most other mammals. Using echolocation, I can hunt insects in the night skies, where variations in light levels make vision less reliable.

Dr. Fenton paints a picture of the hoary bat's sensory world that is both utterly different from ours and surprisingly familiar. Take the example of altering the timing of calls to maximize either the range or the detail. How different is that from looking at the horizon then glancing down to see the time on your watch? Long-range scenic to short-range detail in a moment.

The bat's echoscape has many of the same features as our visual version. The beetle's polished wing covers sound differently than a moth's scaly wings; they *look* different to us. Emitting four pulses a second—and creating four echoes in the same time frame—must create a fragmented, slide-show sort of image of the world, but our vision does something of the same kind: our eyes shift from place to place every fraction of a second, with the brain smoothing together those disparate images to create a seamless "view."

The bat's echolocation is a spotlight sort of sense, but truth be told, human vision isn't that much different—we see details only in a narrow cone in the direction of our gaze. And while it's true that a bat using its low-frequency signals is blind to small insects, let's not forget that we are "blind" to any electromagnetic waves shorter than violet (ultraviolet and beyond) or longer than red (the infrared).

It wouldn't surprise anyone to learn that the hoary bat's brain is laid out in a way that suggests it has wired itself up to make pictures out of sound in the same way that we wire ourselves up to make pictures out of light and shadow.

It is the differences, the notions that seem foreign to us, that make us bump up against Nagel's point that we just can't know what it's like. We can see the colour patterning on a moth's wing, but it's hard to imagine how bats could *echolocate* that. (But do they need to? All they really want to be able to do is approach the moth from behind, swallow the abdomen and cast the wings and head aside.)

How would this ever-changing sonic landscape feel to a bat? Maybe just like vision feels to us. Maybe not.

CHAPTER ELEVEN

Where Did Consciousness Come From?

T HE challenge in this chapter is knowing where to start. (If thinking in circles bothers you, skip immediately to Chapter 12.) Here's one way of approaching the story: We are conscious nowadays. Did consciousness evolve? Maybe. Is there evidence that some of our direct ancestors weren't conscious in the same way we are? If so, it would suggest that consciousness established itself in our brains over time (although the question of whether that would have been a gradual or a sudden process would need to be answered). It is true that a few million years ago, the not-yet-humans living on Earth had much smaller brains than we do (although as tempting as it is to suspect that the two go hand in hand, that's not necessarily true—Flores man, the small-brained but smart fossil recently found in Indonesia, showed that). But to establish itself by evolutionary mechanisms, consciousness—or its earliest, much simpler forms—would have had to confer some reproductive advantage on the animals that had it. Consciousness would have had to be useful.

Maybe it was and is, but there are some heavy-hitting philosophers and scientists who would disagree. They'd say that consciousness is an epiphenomenon, a by-product, something that just happens as neurons go to work but that has no influence over the actual brainwork of actions, words and decisions. The renowned Thomas Huxley, the man best known for his spirited defence of the theory of evolution, set the tone for this line of argument in a nineteenth-century essay in which, ironically, he could find no place for consciousness in human evolution.[1] As far as Huxley was concerned, consciousness has as little influence on us "as the steam whistle which accompanies the work of a locomotive engine is without influence upon its machinery." His contemporary Francis Galton, Darwin's cousin, dismissed consciousness as "a helpless spectator of but a minute fraction of automatic brainwork."[2] While these are vivid images, I prefer the late Julian Jaynes's description of epiphenomenonalism as "the melody that floats from the harp and cannot pluck the strings, the foam struck raging from the river that cannot change its course, the shadow that loyally walks step for step beside the pedestrian, but is quite unable to influence his journey."[3]

Is consciousness useful? After all, is it necessary to experience all the sensations of eating a fresh orange to ensure that you will eat that orange, or would unconscious hunger be enough? The important point is that if consciousness *is* merely an add-on, then it couldn't evolve, because there would be no way for it to contribute to reproductive success. If it doesn't actually *do* anything, it can't help you!

That would still allow for consciousness to get more elaborate as the brain grew over hundreds of millennia, but it would be the bigger brain—and the more and more complex thinking it was capable of—that was being selected, not the consciousness part.*

A book could be written about this debate. To simplify matters, I'm taking the stand that consciousness does have some use and could be subject to the forces of evolution. As William James asked, if consciousness is superfluous, why does it run in parallel with the challenges facing us, wandering when there are no

---

* It is often very difficult to tell what has been selected by evolutionary forces and what hasn't: We use our two legs not only to walk, but also to kick footballs, to dance and to balance on tightropes; but of all those, it was only walking that aided survival and so was honed by natural selection. The other skills just came along for the ride.

crucial decisions to be made but turning up high when there are? Christof Koch and Francis Crick (of DNA fame), who narrowed their search for consciousness to visual awareness, argued that even in this restricted example, consciousness brings the ability to highlight the most salient part of the visual scene, compare it with information experienced before, whether by the individual or by the species, and then communicate the ideal response to those parts of the brain that are capable of bringing it about.[4] They acknowledged that unconscious processing is highly capable but probably unwieldy if decisive action (which of course could be flight or fight, or something much more restrained, like speaking) is necessary; better, they argued, for the brain to come up with one conscious, complicated interpretation, which is then provided to the action parts of the brain for long enough to make such action possible.

It's also true that the brain is a metabolically very expensive organ, consuming 25 per cent of the body's glucose and an uninterrupted supply of oxygen. Consciousness is a significant part of the brain's output, and it's very unlikely that the mechanisms that create it would have been put in place unless they conferred some advantages to the organisms with them. Planning the future, weighing alternatives and imagining what to do all seem to be features of our minds that require consciousness and that could also contribute to reproduction and survival.

How this story unfolds depends a lot on the animals that we are related to and descended from: if they aren't or weren't conscious at all, then it's hard to see how it could suddenly have appeared in us perhaps as recently as a couple of hundred thousand years ago. On the other hand, what if they're all conscious? Then we're nothing special. Obviously this is a much more difficult problem than, say, trying to trace something like the evolution of the hand. In that case, you can work your way back through fossils to find the first animal with a five-digit appendage, then watch as it develops over time until it's a hand, not a paw or a foot. At each step along the way you can try to find evidence of why it changed the way it did. But there's no fossil record of consciousness, just of the brains that might have housed it, and even that is pretty sketchy.

The bottom line is that until we know exactly how many kinds of consciousness there are, when they appeared and in what animals, there's not really much you can say. But you *can* speculate. That's why the science of the evolution of consciousness is some of the coolest science you can do. Cynics would denigrate it by saying it's unconstrained by data, but that does have the virtue of allowing the imagination to run free, which is exactly what's happened here.

Three Thousand Years Ago

Where to start? We're all conscious now, as far as we know, so let's go backward from the present. My main reason for suggesting this is so I can begin this chapter by talking about Julian Jaynes and his one-of-a-kind book *The Origin of Consciousness in the Breakdown of the Bicameral Mind*. This book is an enigma: more people have heard of it, have some idea of what it's all about or have actually read it than you'd ever imagine. At the same time, there isn't a scientist anywhere—at least that I know of—who takes it seriously. This is true now, and it was true in 1976, the year it was published. I'll never forget a prominent consciousness expert saying to me then, "It's a lovely book—I don't believe a word of it." Yet again, Jaynes himself was a great guy—friendly, smart, fun to interview—and his book is beautifully written. Here is the opening of the Introduction, "The Problem of Consciousness":

O, what a world of unseen visions and heard silences, this insubstantial country of the mind! What ineffable essences, these touchless rememberings and unshowable reveries! And the privacy of it all! A secret theater of speechless monologue and prevenient counsel, an invisible mansion of all moods, musings and mysteries, an infinite resort of disappointments and discoveries. A whole kingdom where each of us reigns reclusively alone, questioning what we will, commanding what we can. A hidden hermitage where we may study out the troubled book of what we have done and yet may do. An introcosm that is more myself than anything I can find in a mirror. This consciousness that is myself of selves, that is everything, and yet nothing at all—what is it?

And where did it come from?

And why?

The reader is launched into a world that, by Jaynes's description, is both familiar and strange, and that cries out for explanation. And Jaynes is not shy about providing one. Here is his thesis, in a very small nutshell: until about three thousand years ago, people were not conscious, even though they had fully modern-sized brains. They might have looked like they were, but their brains worked completely differently from ours. Rather than being conscious, thinking about doing things, making decisions, and choosing to act,* they were automatons or, if you prefer, zombies.

These pre-conscious people didn't make up their own minds—they were given instructions. Specifically, their right hemispheres gave their left hemispheres orders in the form of auditory hallucinations: the voices of the gods. When those voices spoke, people acted. The mind was composed of two rooms—it was *bicameral*.

Jaynes's best evidence is Homer's *Iliad*. In it, almost no one acts unless directed to by the gods. There are countless examples. When Agamemnon steals Achilles' mistress, a god warns Achilles not to strike back (and when, near the end of the war, Achilles reminds Agamemnon of this, Agamemnon is in denial: "Not I was the cause of this act, but Zeus"); a god rises out of the sea to console Achilles on the beach; gods start the war and plan the strategy, a god leads the armies into battle, a god speaks to the soldiers at crucial moments. As Jaynes puts it, "the Trojan War was directed by hallucinations."

Jaynes returns to that favourite of all comparisons, the driver

---

* Those of you who bought into Daniel Wegner's thesis about the illusion of free will in Chapter 9 will find this unsurprising.

who is unconsciously guiding the car, to clarify exactly what he means by a bicameral man (or woman). Much of the mind is occupied by the act of driving, responding to changes in the visual scene unfolding before him, hearing the sound of the engine, feeling the road through the steering wheel, totally absorbed by incoming information and responding to it, but never being aware of any of it. Instead, the conscious mind is listening to the radio, or conversing with a passenger or daydreaming. Now, says Jaynes, just subtract that conscious part of it, and you have what it would be like to be a bicameral man.

Jaynes doesn't put all his eggs in the one basket of *The Iliad*. He also argues that the statuary of the time (three to five thousand years ago) was designed in a way to elicit these hallucinations, with prominent, oversized eyes triggering trancelike states that made people receptive to their own hallucinated voices. Temples were the same, designed to focus the attention of worshippers in ways that would enhance their ability to tune into their voices. He ranges over the cultural evidence from several different civilizations to strengthen his argument that the people of the time were split-brained in a different sense from those who are created today by surgery. In those days, the right hemisphere had not even a scrap of awareness, and the left no decision-making power.

What happened to fuse the two? Jaynes claims that growing populations and increasing densities of city life proved to be too much for the bicameral mind, which to function at its best needed a relatively rigid authoritarian social structure, with people of one mind (irony!). The collapse of some civilizations, the resulting hordes of refugees and the inevitable coming together in urban concentrations of people who, although alike in their bicamerality, heard quite different voices, proved to be too much, and the bicameral mind was forced to meld together. Jaynes, well aware that the people of this era already had physically modern brains, is careful to point out that the transformation from the bicameral mind to the conscious mind involved a reorientation or rewiring of the brain, not the creation of new brain stuff, not a wholesale

change of genetics. It was a change of software, not hardware, in what was admittedly some hard-to-imagine process.

Jaynes believed that the environment played such a powerful role in this that if a child from bicameral times were to be brought up in today's world, that child would be just like us, while a modern child thrust back into the ancient city of Ur would be bicameral. It is reminiscent of Sue Savage-Rumbaugh's argument that the linguistic abilities of a chimp or a bonobo depend on the social circumstances, on the way that animal is raised. In both cases, the underlying principle is that the brain is much more than just the wiring determined by its genes; it's also the wiring determined by its environment, and the effects may be not just arrested development (like blindness interrupting the normal development of the visual cortex) but changes to the way people *think*.

If the god-voices were generated in the right hemisphere but heard in the left, how exactly might that have worked? This is a challenge given that the language areas of the brain—for both reception and production—almost always reside in the left side of the brain. Jaynes made a virtue of this difficulty by saying that whatever went on in those areas in the right hemisphere that correspond to the left's language regions must have been extremely important for them not to have been recruited as supplementary language areas to back up the left hemisphere. What could that have been? Why, the god-voices must have been created in precisely those areas in the right hemisphere that, though voiceless, are the analogue of the language areas in the left.*

Having identified the source in the right hemisphere, all Jaynes had left to do was to specify how the hallucinated commands of the gods made their way to the left hemisphere, to be heard and comprehended. He nominated specific tracts of neurons—the anterior commissures—as the ones with the terrible responsibility of carrying civilization on their backs until the arrival of consciousness.

---

* It is also true that in cases where the left hemisphere is damaged early in life, the right is capable of taking on language, although its ability to do so diminishes with age.

I have probably given too much space already to a book that, as I said before, no one active in the field of consciousness actually believes, but I love this book because it makes you think, broadly and imaginatively, about the subject, and in the end that might be all one can ask of a book. But there is one more piece of Jaynes's evidence I must present before the critiques.

Jaynes devotes a large section of the book to schizophrenia, claiming that it represents a holdover, a "partial relapse" to the bicameral days.* It is true that schizophrenics hear voices and that often those voices are giving them commands, which many schizophrenics are inclined to obey. But how exactly would that work? For one thing, the left hemisphere has most of the language skills in the brain. Second, there's good evidence that schizophrenics are hearing their own voices (including one study in which a schizophrenic man wore a microphone, which recorded his own whispering voice uttering the very commands that he shortly afterward reported as hallucinations). Put those two together and the sensible conclusion to make is that the hallucinated voices of schizophrenics are both made up and perceived in the left hemisphere.

That would be the sensible conclusion, but it isn't necessarily the right one. The left *is* language dominant, but our right hemispheres contribute hugely to language production and understanding, especially by generating tone and rhythm, with all their power to influence speech. And there are some brain studies that, while far from being indisputable, suggest that schizophrenics are bicameral. At the very least they show that the right hemisphere seems to play a strong role in hallucinations. In one study of a young male schizophrenic, the right side of the brain, just around the ear, lit up about three seconds before the patient indicated he was experiencing a hallucination.[5] That part of the brain stayed on while other parts gradually came online. Similarly, a two-patient study came to roughly the same conclusions,

---

* It is also true that a surprisingly large fraction of the general population—something like 50 per cent—testifies to having heard voices at one time or another, especially in times of stress—a fleeting moment of bicamerality?

while adding a timeline: the same area on the right side of the brain, together with a partner at the left front, became active a full six to nine seconds before the patients indicated that they were aware of their hallucinations, with other areas, on both right and left, joining them as the hallucination persisted.[6]

What's going on in the right hemisphere in the moments leading up to a hallucination? If these were the heady days immediately after Jaynes's book was published, believers would have been jumping all over such results. Today, the results are interpreted in other, more sober ways, in keeping with the realities of modern psychiatry.

Jaynes's critics—and there were many—attacked this idea on several grounds. They couldn't believe that consciousness could have appeared so recently, that *The Iliad* could be taken so literally, that idols could possibly have been designed to trigger hallucinations . . . I could go on. As far as I'm concerned, it still stands as an intriguing work, an absorbing piece of "software archeology," as Daniel Dennett called it.

Thirty Thousand Years Ago

As I said, the vast majority of people interested in this subject do not believe that modern consciousness came into being just over the last three thousand years or so. But if not, when? Again working backward in time, the next potential milestone is somewhere roughly between thirty and fifty thousand years ago. The more recent date marks the earliest appearance (so far discovered) in Europe of elaborate, beautiful cave paintings, a time when physically modern people had moved into Europe from Africa and the Middle East to compete with the resident Neandertals. In fact, by 30,000 years ago, the Neandertals were gone.*

---

* All these dates are subject to change as new evidence becomes available, as new sites are explored and old prejudices die. There will also forever be artefacts, camp sites and burials that suggest more in the way of consciousness to some than to others, or that the Neandertals were creating their own art objects, something that until recently was believed not to have happened.

Why would cave art represent the onset of consciousness? Psychologists Mark Leary and Nicole Buttermore argue that the explosion of art at this time represents the best available evidence that the highest stage of human consciousness had arrived.[7] Leary and Buttermore create their history of consciousness by linking two different timelines. One is founded on whatever traces of thought and consciousness pre-humans left behind, a tenuous trail through time of stone tools, signs of the use of fire and indirect evidence for foresight, like cooperative hunting. The other, more theoretical, is a five-level classification of consciousness, beginning with a straightforward version that allows an individual animal to be aware of the world around it and to realize that it is the one experiencing that world, through versions of consciousness that permit the possessor to interact with others, then to the ability to imagine the past and future and finally to a full human consciousness in which we wrap all the preceding abilities together and add a symbolic component, identifying others as "moral," "sociopathic" or "responsible."*

These are admittedly arbitrary categorizations of consciousness, but they can make sense of behaviour. For instance, both humans and chimpanzees can envision the future, but chimps' abilities to do that fall short of ours. In the 1920s, Wolfgang Kohler performed a famous set of experiments with chimps in which they had to solve the puzzle of how to get a bunch of bananas hanging from the ceiling. Some figured out that by stacking boxes on top of each other they could reach the bananas, but while they were often pretty good at this kind of reasoning, which involved seeing into the near future, they also in some cases failed to solve the puzzle a second time, suggesting that their ability to imagine

---

* Mark Leary described it this way to me: "Many of our future-related thoughts are very abstract and symbolic—Will I get a good job? Will my children grow up to be happy and well adjusted? Will I be punished for my sins? These sorts of thoughts require the ability to think about abstract qualities of oneself." This is the most complicated form of future thought. Prior to this, people might have envisaged nothing more than the next day's hunt. As soon as minds went abstract, though, people wanted more from their futures, like wealth and popularity. (Yes, these are thought to be advances.)

the future was either limited or inconsistent or both. Kohler himself came to that conclusion.

With this approach, Leary and Buttermore argue that the period technically known as the transition between the middle and upper Paleolithic, but more popularly described by various authors as the "creative explosion" or the "cultural big bang," represents the attainment of full-blown consciousness. It's been argued that there were more changes in human thought in the period from fifty to thirty thousand years ago than there were in the previous 5 million years. Something happened to change a people so different from us that we couldn't imagine what it was like to be them into people just like us. Exactly what happened is up for grabs: was it the human brain bringing together for the first time all its disparate thinking modules, themselves brought into being to accomplish different tasks, or was it the development of symbolic ability—as reflected in the art—or even some amazing mutation? No one knows.

Here's what we do know. First, stone-tool making, the gold standard of evidence for human thought, had already experienced a dramatic revolution. Tools had remained pretty static in design for upward of a million years, but by fifty thousand years ago things had changed. Blades that had been formed by simply flaking one or both sides of a piece of rock had given way to tools obviously designed for special purposes, like chopping or scraping. People had begun to use other materials, like bone, antler and ivory, and tools were elaborated into bows and arrows, spears and fish-hooks, some of them with blades hafted onto shafts. At about the same time, people were on the move, occupying what was then the frozen north, taking boats to Australia, living not just in the caves of popular myth, but in shelters created by stretching hides over poles.

Even as industry was taking off, so were the arts, although they took a little longer. It's possible people had been decorating their bodies with paint or tattoos for millennia, but we will likely never know—we only have their bones. However, this period saw the first definitive evidence of body art, in the form of beads, bracelets, even headdresses, as well as the invention of bone flutes, small carvings of human form, the burial of dead with an apparent eye to an afterlife, and the most dramatic examples of the flowering of culture, the elaborate cave paintings in France and Spain dating back to thirty thousand years ago.* These are stunningly crafted images of animals—the only signs of people being the handprints left behind by the artists—in lifelike poses that attest to great artistic skill, and perhaps much more. Leary and Buttermore contend these were most likely symbolic representations of animals such as bison and deer that were the prime targets of hunters. Isn't that exactly what fully modern human consciousness can do: use symbols to represent real animals, create arrangements of those symbols in out-of-the-way places in ways that must have meant something to them (but are mysterious to us), possibly in an attempt to ensure that a future hunt, executed by the group, is successful? Obviously, we're guessing what the cave art really meant, but it is sure to have meant *something*.

Leary and Buttermore contend that there's a case here for the appearance of fully modern human consciousness at this time, with its symbolism and its capacity for imaging future and past. It's as good an idea as any, and when you take in all the artefacts that these people left behind, it's convincing. But there are at least two major questions left unanswered here: what prompted the final step to full consciousness, and, just as intriguing but much more difficult to answer, what sort of minds did these people really have? If it's practically impossible to imagine the "human" mind prior to this development, is it all that much easier after?

There is one piece of research that addresses this "what kind of minds did they have?" question, and while it is anything but definitive, it is totally fascinating. We have psychologist Nicholas

---

* Archaeologist Steven Mithen argues that these pieces of art represent minds that are capable, for the first time, of creating thoughts and images that don't exist anywhere in the natural world. Rather than holding them in the brain forever, artists recast them as objects—as he put it, "anchors for ideas that have no natural home within the mind."

Humphrey to thank for this. In a classic demonstration of how an open mind can see things that others can't, Humphrey compared ice age art to the drawings of a severely autistic child and concluded that the ice age artists "may actually have had distinctly pre-modern minds, have been little given to symbolic thought, have had no great interest in communication and have been essentially self-taught and untrained. Cave art, so far from being the sign of a new order of mentality, may perhaps better be thought the swan-song of the old."[8]

The autistic artist was a girl named Nadia, born in England in 1967. She was severely handicapped; by the age of six she still hadn't accomplished spoken language and was incapable of normal social behaviour. But by that time, and in fact as early as the age of three, she was an accomplished artist, able to draw animals and people with great skill. Her drawings are extraordinary, especially in that she seems never to have gone through any early stages of artistic development but apparently had this talent visited on her.

Humphrey discovered, many years after many of Nadia's drawings had been published, that there was an uncanny resemblance between her drawings and a variety of animal images on the walls of caves from tens of thousands of years ago. The drawings in both cases are realistic, in both shape and size but also in movement; there is depth and perspective; both sets of drawings include many animals that have been superimposed on others. In the case of the ice age artists, this has been interpreted as an effort to depict numbers or movement. There are also animals that appear to be a combination of two, drawn as if the partial completion of one became the starting point of the next, the artist having changed his (or her) mind in mid-flight.

The resemblances are extraordinary, but the lessons aren't as clear. Humphrey admits that perhaps there's nothing to be taken from what could be simply an amazing coincidence, but such caution aside, he makes two points: one, that he is not implying by this comparison that he thinks ice age artists were autistic, or

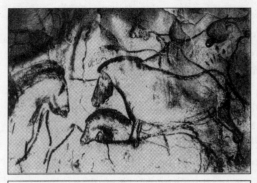

These images say more about two very different, yet very similar minds than any amount of verbal analysis could. The first and third are from the ice age caves at Lascaux; the second and fourth are drawings by Nadia, the young autistic artist.

that Nadia had an ice age brain, but, two, that the comparison does tell us something about what we can't assume about ice age artists, and that is that they had language. Nadia didn't, and she was capable of art that, at least in Humphrey's mind, is equivalent to that produced by the ice age artists.

There is even evidence from Nadia and other autistic artists that language is in some way inhibitory to their artistic expression. When Nadia reached the age of eight and began, with intense language training, to acquire a larger vocabulary, the quality of her art declined. Humphrey picks up on this to argue that it supports his contention that the ice age artists were the last of the old mentality *because they didn't have language*.

This whole proposal is Jaynesian in the sense that most commentators disagree—sometimes forcefully—with Humphrey. They have many reasons: Nadia's art and the ice age art may look superficially similar, but they are, at a deeper level, very different; that this is a classic case of n=1, and that Nadia, by herself, proves nothing; that there were, at the same time as the ice age paintings, other art forms flourishing around the world, and it's hard to believe either that no human was speaking by this relatively recent epoch or, conversely, that somehow the ice age artists were the only modern humans not speaking (although that is something that Humphrey muses about). The best most can bring themselves to admit is that Humphrey (again reminiscent of Julian Jaynes) deserves credit for shattering the calm of paleoanthropology. For me the value in this approach is that it forces us to think about what people far removed from us in time and space might have been like. It's Thomas Nagel's old question rephrased once again: "What was it like to be an ice age artist?"

I can't leave this epoch of human life without addressing the age-old question of what happened to the Neandertals. The Neandertals were, in the minds of most experts, a side-branch of the human lineage, one that coexisted in some locations for thousands of years with modern people yet apparently contributed little if any genetic material to us (although the idea that humans and humanlike creatures wouldn't have at least cast sideways glances at each other seems preposterous).

There is no agreement yet on why the Neandertals, who survived beautifully through dreadful ice age conditions in Europe prior to the arrival of modern people, would then go extinct within a few thousand years. Were they not smart enough, not adaptable enough to the changing environment of Europe thirty thousand years ago? Were their language skills not up to the task of survival when it was complicated by competition with new, smart arrivals? There is no universally accepted answer, but one thing is clear: if it would be difficult for us to inhabit the minds of our own direct ancestors thirty or forty thousand years ago, how much more difficult would it have been to know what it was like to be Neandertal?

Scientists are sometimes shy about speculating about what it would have been like to be Neandertal, but several novelists have taken a crack at that feeling, and while Jean Auel is the best known among them, for her series that began with *Clan of the Cave Bear*, British novelist William Golding wrote a much earlier book, in the 1950s, called *The Inheritors*. And while Golding's take on the Neandertals has been dismissed as being old-fashioned, in that he sees brutal, head-to-head combat as the cause of the Neandertals' disappearance, at the same time he captures what could have been some of the mental differences. His Neandertals are just able to translate thought into halting speech, but that process is confused by the leftovers of a consciousness driven by vivid mental pictures, some of which can be called up, some of which arrive unbidden. A mentality driven by pictures rather than symbols . . . it might have been, although I can't imagine how we will ever know.

This chapter might seem a bit flaky, being an account so far of what is likely fiction masquerading as science—together with admitted fiction—but is this so different from the mainstream scientific accounts of the evolution of consciousness? It's all speculation, every investigation indirect (although to be fair, some might be somewhat more constrained by the available data than the rest).

Three Hundred Thousand Years Ago?

There's a question mark after "Three Hundred Thousand Years Ago" for a good reason: there isn't much point in continuing this trip back in consciousness time, for the same reason that tracking footprints in a blinding snowstorm quickly becomes hopeless— the trail is simply too indistinct to make sense. Before fifty thousand years ago, the absence of agreed-upon art or any other cultural evidence makes inferences about the appearance of consciousness, and assigning a time to it, much more difficult. Nonetheless, there are pieces of evidence that need to be considered, and the murk of the past doesn't mean that speculation about the value of consciousness, the importance for reproduction that it would have had to have to be selected by evolution, is irrelevant. Most suspect there's something about consciousness that makes it valuable for us, and that something must have made an appearance sometime, whether that time can be pinned down or not. Three hundred thousand years ago is as good a candidate as any, because it is about at that time that stone-tool making suddenly took off, emerging from a period of millions of years of little, if any, invention or innovation.

Steven Mithen, a paleoanthropologist at Reading University in England, agrees that while the appearance of art fifty thousand years ago represented a dramatic—the *most* dramatic—step forward in human consciousness, there must have been an earlier scene-setting process, a not-yet-mature consciousness that evolved around the making of stone tools, especially handaxes.[9] He puts a twist on the traditional thinking here, by saying that not only does the manufacture of such tools suggest what might have been going on in the maker's mind, but that such tools encourage, in return, new ways of thinking on the toolmaker's behalf. Tool making is a two-way street, at least for the brain. Handaxes appeared early in the human fossil record, about a million and a half years ago (although cruder stone tools had been made for at least a million years before that) but then remained pretty much unaltered for the next million years. There were apparently no changes made to make them work or look better.*

Mithen proposes that there are four mental abilities that have to be in place before anyone can make a handaxe, and he's convinced all were in place long before the first ones appear to have been made. These qualities are precise coordination, an understanding of how stone is likely to fracture, planning and a sense of symmetry. Without fine motor control, it's impossible to strike stones at just the right angle and with just the right force to dislodge the size and shape of flake you want. That plays into the understanding of the stone itself, an understanding that even chimpanzees have in cracking nuts with rocks: with practice they are able to open the shell without damaging the nut inside.**

The importance of planning extends beyond looking ahead to the strikes you want to make on the stone. It also involves transporting the handaxes to the place where they will be used, and here again chimps have demonstrated the same sort of previewing the future. Mithen chooses the sense of symmetry as his fourth key characteristic not just because an evenly balanced tool would likely be a better tool, but because an attraction to symmetry—especially in the bodies of potential mates—is a very ancient mental predilection. Each of these four requirements is likely to have evolved independently, and earlier, than tool making but were brought together into a kind of short-lived tool-making consciousness.

---

* Buttermore and Leary think that the reason that stone tools didn't evolve much over hundreds of thousands of years is that the toolmakers had no robust concept of a future. Without that, there's no perceived need to improve things to be able to do things better in the years or months to come. The fact that these tools varied little, not only through time, but across space as well, is a stunning demonstration of the lack of individual culture—even chimpanzees design and use tools differently from one part of Africa to the other. That doesn't mean that these proto-humans were dumber than chimps, but they were undoubtedly different from us.
** It's worth noting that, contrary to our males=tools thinking, female chimps are more persistent and skilful in their tool use than males, suggesting that females might have played the leading role in tool innovation and, by extension, the development of consciousness.

But what brought them together? Mithen suggests it was talking to oneself—"private mutterings accompanying the crack of stone against stone," as he puts it, an out-loud version of inner speech, at least in this example. According to Mithen, when many toolmakers coached themselves in handaxe making, this speech brought together the otherwise disparate mental qualities (which they already had) into a tool-making consciousness—not a consciousness that attained the heights of the cave painters, but a new level of awareness nonetheless. It would have been the sort of "software" change that Julian Jaynes envisioned, although in this case not yet producing the fullest human consciousness.

Mithen isn't the only one who sees language as the main force behind the development of consciousness. Because the emergence of language is pegged at somewhere between 100,000 and 200,000 years ago, roughly at the time of the earliest fossil remains of people with modern brains, it seems that the whole project could have come together at one time. But even Mithen's description of the role language played is a little vague—what exactly could the ability to speak do? Some argue that it is the one thing that allows us to reflect, not just on our situation, but on the fact that we're reflecting on it. No one is sure about dogs, cats or chimps, but most consciousness researchers suspect we're the only ones on Earth who can think about thinking, and for most of us speech is essential to that ability.

Scientists who think about the origins of language can catch you by surprise by making unexpected links to other human behaviours, based on the fact that both employ carefully thought out sequences of movements. We would be unlikely to stumble on those associations today, fully equipped with language as we are, but to think about origins you have to transport yourself back to a time when our ancestors weren't really fully human, and that requires an open mind.

Derek Bickerton is a linguist who's thought long and hard about the appearance of language in our past, and he thinks that language and consciousness may be connected indirectly, by an as-yet-unknown third party. This suggestion gets around the chicken-and-egg problem of whether consciousness begat language or vice versa, because it allows the possibility that both were enabled by something else. And Bickerton suggests that that something else could have been timing (it is, after all, everything).

The World Around Us

Psychologist Bjorn Merker of Uppsala University in Sweden has come up with a novel suggestion for why consciousness just had to evolve.[10] He argues that it all comes down to mobility: as soon as animals started to move, they needed to be conscious. Why? As soon as they became mobile, they needed to devote a significant amount of brainpower to the prevention of confusion caused by their own senses.

When you perform an act as simple as walking from one end of a room to the other, looking from side to side, the room appears to be a steady, immobile place, an unchanging space through which you are moving. But think about the sensory information that allows you to create this picture. Your eyes are darting back and forth, your head swivels from side to side, your body sways forward and back and side to side as you move—your brain is being buffeted by a bizarre mish-mash of visual input. As Merker puts it, "A single eye movement sets the world sweeping rapidly across the retina in the opposite direction, yet the world we experience undergoes not so much as a tremor on its account."

The same is true (although to a lesser extent given our devotion to the visual) of the sounds you hear, the feeling of the floor under your feet, even the smell of last night's chili emanating from the kitchen. It would be totally unhelpful for you to be aware of this cacophony of the senses. It would be disruptive. You want the room to appear as a room, not a Disney ride. Your brain deals with this problem by relegating most of the raw data of this incoming sensory stream to the unconscious, where it is massaged, edited and processed into a form that makes sense.

This is only the first half of the story. We use that sensory information to take action, sometimes dramatic action, like leaping to safety from an attacking sabre-toothed tiger, sometimes not, as when we simply whisper to a companion. The incoming leads to the outgoing, admittedly with some thinking in between. Again in this case, the complex processing that turns mental activity into action is something we need not be aware of, and so it can reside in our unconscious.

Merker argues that for mobile animals to be effective, and stay alive, it made evolutionary sense to bury all the mental processing in the unconscious, leaving only the final product available to awareness. We need to know, we need to react; we just don't need to know how we do it.

It's all about the fact that we inhabit bodies—they are our reference point for the world. It is literally the world around us, and we need to know how our body is relating, moment by moment, to that world.

I couldn't help but think some of the fantastic robots I had seen in Japan when I read Merker's paper. If the assembly line robot soon gives way to the autonomous robot that moves about the landscape, will such robots need consciousness? Or more intriguing, will they, when they have the computational wherewithal to navigate through the space around them, develop consciousness spontaneously?

Ancient humans were hunter-gatherers, and there's some evidence to suggest that throwing stone axes or blades would have been one of the major hunting techniques. Although I have had an angry male chimpanzee throw stones at me—and hit me—I wasn't impressed with his accuracy. He was lucky. On the other hand, it appears that we have an uncanny knack for accurate throwing, something that likely evolved over tens of thousands of years. But throwing something at a moving target—or a stationary one, for that matter—requires exquisite control of timing. The error of releasing a stone a fraction of a second too soon or too late is multiplied many times over the distance the weapon travels. Bickerton makes the point that this sort of precision timing requires hordes of neurons acting in synchrony. A handful just couldn't ensure the necessary accuracy. The demand for bigger neuron numbers could have paved the way for a bigger brain, and once that was in place, other, unanticipated possibilities could arise.

For instance, refining the sequence of movements for accurately throwing a projectile requires coordinating every single joint movement, from hip turn to shoulder rotation to elbow straightening to wrist flip to the uncurling of the fingers, and more important, ensuring that each movement is predicated on the one before: for instance, your wrist doesn't move until your forearm is acclerating. Bickerton (and others) see in this precise sequence the movement analogue of language, in which subject, verb and object each depends on the positions of the other in a sentence. (Others have seen the same resemblance in the step-wise manufacture of stone tools.)

Throwing then becomes a kind of training ground for the brain to acquire the sequencing of thoughts in language and the tongue and lip movements of speech. The area of the brain necessary for structuring speech, Broca's area, overlaps with the areas for coordinating the arm movements of throwing, or even the finger-joint movements necessary for threading a needle. It's all about movement: lips, tongue and breathing for speech, arms and hands for throwing.

A later version of the same thing has been advocated by Steven Mithen, who sees the explosion of art and culture as the result of many different areas of the brain, with their own specialized abilities, coming together to provide a new set of mental abilities, possibly including full-blown consciousness.

So, in a way that is not yet clear, tool making, throwing, speaking, planning and reflecting may have been intertwined in a way that, although producing little evidence of anything for well over a million years, was likely working its way through the evolving human brain. It's hard to assign dates to any of this, but some time beginning around 300,000 years ago, things started to accelerate, and by 50,000 years ago, change being wrought by a new kind of thinking was dominant.

Pinning the emergence of consciousness down to a particular time is, let's face it, pretty much impossible. And there is still the question of why. What does consciousness do for us that couldn't be done by the unconscious mind? You could guess that some of the features already identified as steps towards consciousness, such as planning and being able to imagine the future, would enhance survival, and if these were enabled by consciousness and consciousness only, then that would be why it evolved. Those who couldn't plan ahead—perhaps the Neandertals—failed; those who could, did.

But there are plenty of other ideas. One that I like was put forward by Matt Rossano, who, while admitting that comparative studies of human consciousness are impossible because there's only one remaining species of our lineage, argues that *expertise* is the key to consciousness, and a study of it can help answer the question, why are we conscious?[11]

Rossano concentrates on that part of consciousness tied to attention. While I've been writing this book, I've been examining and re-examining that most popular of consciousness experiences, being unaware of the last ten kilometres of highway that you've driven. At first I thought that the main reason we pay no attention to driving is that most of the time it's really boring, not worth wasting consciousness on. But that's only partly true: I've also noticed that the more difficult and challenging the traffic is, the more focused my consciousness is on the road. In other words, consciousness is recruited when it's really needed. This ties in with the idea that the job of consciousness is to preside over learning but that once a routine of any kind is learned, consciousness moves on—it seeks novelty or situations in which routine learned behaviour may not be enough.

Expertise is the thing we acquire when we use our consciousness to learn something. Driving would actually be a perfect example. Once you've driven for a while, you don't mentally rehearse the sequence of actions you take immediately every time you sit behind the wheel. You just do them. But all the evidence points to the fact that to acquire expertise you need to practise. There's even something called the "ten-year rule" which claims that true expertise requires at least ten years of preparation, ten years of rigorous practice. One study showed that the best musicians had put in ten thousand hours of practice before they were twenty. Expertise in this case refers to the absolute highest levels: being a scratch golfer, say, or a tournament chess player. Being the best on your block doesn't count.

Rossano believes that only humans do this: animals for the most part acquire survival expertise by play, not by deliberate, repetitive practice. Acquiring expertise does require some deep thinking—at all times you are evaluating yourself against those with more skill than you have, and implied in that is the idea that you are always looking forward, changing and bettering what you're doing. Of course, what immediately comes to mind is that drought of innovation in stone tools that persisted for a million years plus. Where were the expertise-seekers then?

Although I've been referring here to expertise, deliberate practice and planning, Rossano contends that they all require consciousness, so if you find evidence of expertise, you've found the footprint of consciousness. He cites an intriguing study to back that point of view, one that in some ways resembles my experience of using consciousness on the highway when things get tight. In the study, expert pianists had to memorize a short piece, but then, when asked to play it back, had to make unforeseen changes, like playing every other bar or playing with only one hand. Nonetheless, their performance was virtually unhampered, suggesting to the experimenters that the pianists couldn't possibly be using well-rehearsed sequences of actions that might be unconscious but were actually relying on the flexibility of consciousness.

Rossano is concerned about the possibility that expertise could be acquired unconsciously, but he cites evidence that unconscious processing cannot achieve long-lasting routines, combine things in unusual ways or support one's intention to do something, all three of which are involved in acquiring expertise. There are actually many studies that show that being able to adjust behaviour quickly to changing circumstances requires consciousness.

This is all great, but is it any easier to go back in time and identify the appearance of expertise than it is to identify the presence of consciousness? Maybe. Most who have studied animals repeating some action over and over have come to the conclusion that they, like kittens repeatedly batting a tinfoil ball, are really playing, not deliberately practising. Another issue is the true nature of practice: humans try to perfect whatever it is they are practising by making tiny changes in the routine and then checking out the results over and over, but animals don't do that—that captive chimp that hit me with the stone never spends any time practising his throwing.

When Rossano looks to the past, he sees some evidence of expertise in the making of stone tools. He agrees with Steven Mithen that the production of handaxes must have required time, energy and facility in design, but again, they didn't change much for a million years, not exactly a testament to increasing expertise.* Rossano argues that if these early humans had really been practising, we would find a range of quality in the products, but instead the evidence suggests that the best handaxes came from the best materials, with the maker's skill contributing much less.

However, by about 300,000 years ago or less, tool making took off, with complex implements featuring more than one part appearing for the first time. It's at this point that some experts see a definite link to language: tools that require a series of steps, each one dependent on the one before, are like sentences, with each noun or clause standing in the same sort of relationship to each other. It seems a strange idea to us now, recognizing a similarity between throwing weapons, or assembling tools and speaking, but it could well have happened that way.

When you put all the available evidence together, the scenario

---

* Nonetheless, the handaxe issue is an interesting one because there are sites where hundreds of handaxes have been found, none of which appear to have been used. Who was responsible for this oversupply of handaxes, and what were they for? Some have suggested that some really skilled handaxe makers were making them not to be used, but to impress women. Funny, that assumes the makers were male.

that is created is one in which full, modern human consciousness likely emerged gradually as an accompaniment of changes in the human brain, although those changes were not restricted to increases in size but were likely to have been wiring and organizational changes triggered by social and environmental change. The consensus right now seems to be that the earliest hominids, such as the Australopithecines, were likely not fully conscious. At least they left us no evidence that they were. It's not until the appearance of art that most paleoanthropologists are willing to concede that consciousness was upon us. Unfortunately, between the two there is a grey area big enough to drive a truck full of handaxes through. And that may be the way it forever remains.

CHAPTER TWELVE

The Emergence of Consciousness

You can make a good case for the fact that consciousness made its appearance, likely gradually, during the descent of modern humans from our hominid ancestors. Is it also true that consciousness establishes itself, gradually, as we develop from infancy to adulthood? Or are we aware from the very beginning? There is reason to believe that gradual is correct, but establishing the emergence of consciousness in a human life is plagued by the same problem as the evolution of consciousness: a shortage of unambiguous evidence.

The roadblock is the same as the one we encounter when trying to figure out if animals are conscious: with the possible exception of a handful of great apes, we aren't able to quiz animals. Children generally don't talk for at least the first two years of their life, and until then scientists are forced to resort to indirect means to try to establish if those children are conscious, and if so, to what degree. Are they in the process of developing full adult human consciousness, or are they already there?

But wait a minute! Am I not overlooking something crucial? We were once infants ourselves and so have access to the memories of that time, a source of direct, personal information about early consciousness. So we can simply try to remember whether at the age of six months, one year or two we actually experienced the world in the same way we do now, as thinking conscious beings. By doing that we would certainly have an advantage over those trying to establish the presence of animal consciousness.* However, getting access to those memories is another matter entirely. It sounds like a good approach, but it hits a memory wall, called childhood or "infantile" amnesia.

The term refers to the fact that we are unable to remember anything from the first three or four years of our lives.** That amnesia doesn't imply that we have no memory up to that age — quite the contrary. It's easy to demonstrate that children younger than two can remember things they've done in the months before (some have even been shown to recall events that happened when they were only six months old), but once they pass that mysterious boundary, those early memories appear to be lost forever. You may feel strongly that you have kept them, but don't forget that parents' and grandparents' stories about what you did, reinforced by family albums, can plant those memories, and there's tons of evidence that you cannot reliably distinguish between those and authentic memories derived directly from your experience.

The cause of childhood amnesia is controversial, but as the evidence slowly accumulates it begins to look as if it's not so much a case of the door slamming shut on a collection of early memories as that the ability to create long-lasting memories is established only gradually, with several different influences playing key roles. It is remarkable, isn't it, that we accept, without a second thought, the extraordinary fact that our lives are marked by an impenetrable memory barrier, forever erasing the early years of

---

* I suppose there are people who claim that in a previous life they were some sort of animal, and so could tell us what that was like, but those reports haven't made it into the consciousness literature.

**There are claims to the contrary. In a radio series on the brain called *Cranial Pursuits* that I worked on for CBC radio in 1993, journalist Chris Grosskurth interviewed a Washington, DC, maître d' named Jacques Scarella who not only had a phenomenal memory for who had eaten in his restaurant, when and what they had ordered, but also claimed to be able to remember events from the first few months of his life. Such claims are unfortunately very difficult to corroborate independently.

our lives in a way that can only be compared to the complete pre-accident amnesia of people suffering serious brain damage.

How does this connect to the establishment of consciousness? To argue that children don't start life fully conscious—something you might suspect from a newborn's behaviour—you have to accept that consciousness comes in different flavours, and that children less than two years old are conscious, yes, but in a less elaborate form than they will eventually develop. Newborns aren't likely thinking about their futures, but they probably have a kind of consciousness, one that connects directly to the senses and allows the experience of the taste of food, the shiver of cold, the pleasure of being next to the warmth of mother, without thinking any more about it. It would be awareness of the senses but that's all. However, things would be different once a sense of self is established: not just taste, but "this is one of the best oranges I've ever tasted"; not just cold, but "the last time I was this cold was when I was in Edmonton"; not just the warmth of mother, but "I like being held by my mother better than being held by my babysitter."

Childhood amnesia also connects in that an important quality of consciousness is the awareness that you are the same person today that has existed before; the catalogue of experiences you've collected all happened to you. Without that personal history, without the realization that your life extends both backward and forward in time, your consciousness is incomplete. In that sense, the emergence from childhood amnesia represents the establishment of consciousness.

We are at an interesting point in the understanding of childhood amnesia: some of the possible causes have been dismissed and a few crucial influences on the development of long-lasting personal memories have been identified. One of the older ideas that I already mentioned was the theory that we had no memory prior to the age of three or four. Today there is no doubt that prior to that barrier—or, putting it more dramatically, during the amnesia period—we have a perfectly functioning memory. Psychologist

Katherine Nelson of the City University of New York cites a frag-ment of so-called crib talk by a child named Emily when she was thirty-two months old. Emily was being taped, without her knowl-edge, while lying in her crib after her parents had said good night and had left the room. The tone of her comments apparently changed dramatically after her parents left—she switched from anxious efforts to persuade them to stay to reflecting on the recent events of her life: "We *bought* a baby, cause, the well because, when she, well, we *thought* it was for Christmas, but *when* we went to the s-s-store we didn't have our jacket on, but I saw some dolly, and I yelled at my mother and said I want one of those dolly. So after we were finished with the store, we went over to the dolly and she *bought* me one. So I have one." [1]

There's no doubt Emily could remember what had been hap-pening to her, and in many ways those memories were organized in the same ways ours would be. Other studies have established that children under the age of four can remember events that happened as much as six months earlier. However, when the texts of Emily's discussions with herself were analyzed, they showed, as does this example, that most of the memories she expressed were concerned with unremarkable daily activities, omitting what you'd expect would have been the earthshaking events of her life, like the birth of her baby brother or trips to see relatives. Of course, in the end she forgot both the significant stories and the trivia of her daily life.

Freud, not surprisingly, attributed the loss of those memories to repression (he introduced the term "infantile" amnesia). In his view, these are memories from the oedipal period and are just too hard to handle, too emotional to be allowed to enter conscious-ness. However, researchers have shown that these early child-hood memories are no more threatening or uncomfortable than later ones—Emily's dialogue with herself being an example—and the Freudian view doesn't play a major role in scientific thinking about childhood amnesia and consciousness.

Nor is it the case that memories from infancy can't be dredged up because they happened too long ago. It's true that twenty-year-olds can't remember when they were two, but eighty-year-olds can remember things that happened to them when they were ten, so the reach of memory can't be to blame. Again, it's not so much losing memories that seems to be crucial, but gaining the ability to remember.

The factors that do seem to be involved are the development of language, mother-child communications and the establishment of a sense of "self," and it may even be possible to link them all to some crucial milestones in brain development. It's important here to acknowledge that there are different kinds of memory, and it is only what most refer to as "autobiographical" memory that is important here. This is the "I was there and did that" sort of memory, not memories for physical skills, such as riding a bike, or general knowledge (that four-legged animal is a dog, and this one is a cat).

Even though young children are perfectly capable of remembering things that happened to them months before, there are signs that those memories are not acquired or retained in the same way as they are in adulthood. In tests where children are required to imitate a series of actions demonstrated to them by adults, the older the child, the longer he or she can remember what to do. So, for instance, one-year-olds can remember a correct sequence only for about a month, but twenty-month-olds can remember for up to a year. As children reach the age of two, they are accomplished at such routines, including games like peek-a-boo and feeding or bathing a doll. But none of these can be taken as evidence that an autobiography has been established. For instance, eighteen-month-olds do refer to the past, but in a fleeting, piecemeal way, and usually in reference to something that has happened just minutes before.

Language is enormously important: as children become more skilled with language, their ability to describe events in their own past grows. The dependence of memory on language has been demonstrated most vividly by experiments in which the ability of children to remember a game has been compared with their vocabulary. In one study, by Harlene Hayne and Gabrielle Simcock, children between the ages of two and three were introduced to a game called "The Magic Shrinking Machine."[2] A child put a large toy into the machine, then turned it on. Lights started flashing, the experimenter turned a handle that produced a series of strange sounds, and then the child opened a door at the front of the machine and found an identical, but much smaller toy. It must have been pretty exciting. At the same time, the child's vocabulary was assessed. Then these children were tested six or twelve months later for their memory of the machine. The results were astonishing.

In every case, the child being tested described the game using vocabulary he or she had had at the time of the first test, six months or a year earlier. The younger the child at the time of playing the game, the less complex the vocabulary, and the *less detailed* that child's verbal memories. However, at the same time, children were able to recall additional facts nonverbally, by identifying pictures of the toys and replaying the actual game. So they actually possessed more detailed memories; they just couldn't express them with words. Their memory for the game was clearly limited by the verbal skills they had when the game was played, regardless of how much more accomplished linguistically they might have become by the time of remembering. Admittedly, this is only over a time span of one year, but the implications are clear: detailed memories that we can express verbally—the kind that we can retain into adulthood—are not formed until a child acquires language.

There's something eerie about this: we all must have had similar experiences when we were two or three years old, but as vivid as they might have been at the time, unless we could revisit them with pictures or actually go back to the place where they happened, our memory for those events is dependent on one simple thing: did we have the words to describe them? No words, no memory.

Might there be wordless memories sitting there somewhere in our minds? Maybe, but your chances of retrieving them are close to zero: even if you re-create the moment, the fact that those memories have had no opportunity to be revisited and refreshed makes their retrieval highly unlikely. Or you might have a shadowy recall, but one that has been modified and updated so many times that its true origin has been lost forever. The idea that prior to the age of four we likely didn't have the necessary vocabulary and so can't remember anything from that time is unsettling, partly because once we've established an autobiographical memory, we want it to be complete, partly because we think of memories as existing on their own, independent of what we can say about them. Apparently that is not the case.*

There are several things happening here: not only is a child becoming more skilled with language, and so is able to add more detail to the descriptions of experiences, but language itself appears to be helping to organize those memories. Some researchers like to refer to the emerging language as a scaffold, on which the elements of a story that will be incorporated into memory are built. It is also true that the greater the language skills, the easier it is to share experiences with others, and that in turn helps create fuller, more detailed memories. The role of others is critical: not only do caregivers, especially mothers, help pry more detail from children's memories by asking them for more, but even the mother's style of reminiscing is, maybe surprisingly, important in determining the development of her child's memory.

Those mothers who are, in the jargon, highly elaborative in discussing past events with their children—who spend time filling in the details of the past with their children and encourage them to describe their experiences more fully—raise children whose first memories are earlier and who can remember more about those experiences. Children whose mother's side of the conversation is

---

* There are rare cases of children who have not acquired language until very late in life. The most famous is Genie, a girl who was terribly emotionally abused—including not being spoken to—by her parents. Only when she was removed from their care when she was thirteen did she begin to learn to speak, but as far as I know there is no information about her early memories, and given her circumstances, it wouldn't be surprising if such memories, had they existed, would have been too painful to access anyway.

simpler and more repetitive, whose mothers are content simply to repeat the same questions over and over when sharing a memory, don't perform as well on tests of early memory.

Here are examples that make the distinction clear,[3] first a conversation between a child and her highly elaborative mother:

Mother: What was near the ocean that you played with?
Child: I don't know.
Mother: Do you remember that we used to walk, we used to walk on the beach. . . .
Child: Um hmmm, Mommy.
Mother: And what did we pick up?
Child: I don't know.
Mother: You don't remember?
Child: You tell me.
Mother: Remember we picked up sea . . .
Child: Uh huh.

Compare that to this, an example of a mother who doesn't elaborate:

Mother: Who else went with us? Think about who was in the car when we went. . . .
Child: Tyler.
Mother: Did Tyler go with us?
Child: Yeah.
Mother: No, Tyler didn't go with us. Who else went? Did Daddy go?
Child: Yeah.
Mother: He did? Now think about who was in the car the day we went.
Child: You and Daddy did.
Mother: Daddy wasn't there.

It's really all about storytelling. Elaborative mothers are filling in helpful details when they expand a child's story, but they're doing it in an ordered, narrative way. Their child can't help but learn that recalling the past is a matter of remembering and telling stories with a beginning, middle and end. The child isn't passive either: even though the mother does most of the work in the beginning months, the children in these relationships soon learn to exert their share of control by giving the mother positive feedback; they are encouraging the very maternal behaviour that is good for them.

Imagine that: the way your mother told you stories and quizzed you about experiences you'd shared enhanced your memory of those events, helped you figure out some sort of script for events like this so you'd remember them better, and likely established the date of your earliest memory—a perfect example of your social life guiding the development of your brain.

Language, social life and the combination of the two are obviously hugely important for a child's ability to remember important events (and for the establishment of adult consciousness). But they aren't the only factors: remembering that the memories we're talking about here are autobiographical and that such memories can exist only when you *have* an autobiography, then the knowledge that you are a unique individual, that you are the same person who is centre stage in the events of your life, that you are likely going to continue into the future, that you are different from your siblings, parents and friends—that sense of "self" is also crucial.

But again there's a problem. How exactly do you find out if a young child has that self-knowledge, especially if that child's language hasn't progressed far enough to be able to answer that question? There is a way: the mirror test, first mentioned in Chapter 10. This is the test devised by Gordon Gallup Jr. nearly forty years ago to determine whether chimpanzees knew themselves. Gallup anaesthetized the chimps briefly, put a coloured mark on their foreheads and, when they awoke and looked in a mirror, the ones who had been marked touched their own forehead repeatedly. They had passed the mirror-recognition test, suggesting that they had a sense of self.

Gallup used these results to argue that chimps and other great apes are conscious. The same mirror test has been used with children and most pass it by the age of one and a half or two. A more elaborate version of the mirror test incorporates a delay, testing whether a child has not only a sense of self, but a time sense: knowing that the self now is the same as the one in the past. In this test, as three- and four-year-old children are occupied by a game of sorting cards, a sticker is put surreptitiously on each child's forehead. Then, a few minutes later, the children watch a video of what they've just been doing and point to themselves when they appear in the video. Three-year-olds do that and nothing more, but most four-year-olds reach up and try to remove the sticker from their forehead. So by the age of four, children appear not only to have a sense of self, but to be aware that the self has a history.

One of the final steps towards consciousness is what's called theory of mind (also mentioned in Chapter 10). This is the idea that I can infer what's in your mind and that I'm aware that you might be thinking different thoughts than I am. I know that you have your own mind. An important part of theory of mind is the understanding that you (or I, for that matter) might be thinking false thoughts. But again there is evidence that theory of mind doesn't just suddenly arrive in a child's brain. Michael Lewis of the Robert Wood Johnson Medical School in New Jersey argues that there are at least four stages of acquiring theory of mind: the first level is *I know*, the only one experienced by children until they are about two. This is superseded by level two, *I know I know*; children are now aware of their knowledge and can turn their attention inward and figure out what they know. "Ah, I remember that!" is an expression of knowing that you know.

The next level is *I know you know*. This is practically all the way to theory of mind, because at this level children know that other people have their own minds. This stage marks the beginnings of deception, because children can see that they know something that *you* might not know. And, finally, level four, according to

Lewis, is *I know you know I know*. This is where complicated human behaviour, of the best and worst kinds, emerges. How many movies use the *I know you know I know* principle as a key plot development?

Lewis places the attainment of full theory of mind at about the age of four, around the time when the curtain of childhood amnesia falls. Coincidence? It is true that memories about events in one's life can make sense only if there is a life to look back on, a full mental life with all the features of adulthood, including theory of mind.

And what about that brain development? Is there anything there that correlates with this dramatic change in a child's conscious world? There might be. Studies of brain development have shown that at about age four, there's a reversal of the way things have been going up to that point. Until that age, growth has dominated, with the number of neurons increasing rapidly, causing the entire brain to expand. But at four there's a crucial change: the population of neurons actually starts to fall, and the white matter, the insulating sheathing on the axons that allows neurons to communicate with each other, starts to increase. So the brain is being pruned, especially the frontal lobes, even while communication is being enhanced. Some researchers suspect that this enhanced communication among neurons facilitates the cross-referencing of information necessary for autobiographical memory and consciousness. So far, about all that can be said is that this thinning of brain cells happens at about the same time as the beginnings of long-lasting autobiographical memory; there's no evidence yet of cause and effect.

However the data are parsed, one thing seems clear: nothing so far suggests that childhood amnesia is suddenly relieved by some abrupt change in a child's life. In fact, the reverse seems true— the ability to create the kinds of personal memories that first appear at about age four is put in place gradually over two years or more of social and brain development. Even when childhood amnesia ends, it doesn't end abruptly. One study showed that there is on average a gap of one year between an adult's earliest and next earliest memories.

It's hard to keep track of these diverse influences on the development of consciousness, but psychologist Nicholas Humphrey, whose discovery of the similarities between the art produced by the autistic girl Nadia and the ice age artists I described in Chapter 11, has applied his creative talents to the emergence of consciousness, by likening his infant son's brain to that of the symphony orchestra arriving on stage and beginning to warm up. The individual players and their instruments are, at first, on their own, each making sounds unrelated to the others; all is dissonant. Occasionally two or three horn players or a couple of violists play together, but that congruence is lost in the din of preparation. The concertmaster arrives, and suddenly the orchestra is together, tuning but not yet playing. That requires the presence of the conductor, who, while not actually adding any sound to the mix, ensures that out of what was cacophony emerges the cooperative complexity of a musical composition. All are attuned to the project at hand. Humphrey imagines the same sort of process going on in his son's mind:

There he is, thrashing about. The difference between him and me is precisely that he has as yet no common project to unite the selves within him. . . . Even as I watch, however, I can see things changing. I realize the baby boy is beginning to come together. Already there are hints of small collaborative projects getting under way: his eyes and hands are working together, his face and voice, his mouth and his tummy. As time goes by, some of these mini-projects will succeed; others will be abandoned. But inexorably over days and weeks and months he will become one coordinated, centrally conscious human being."[4]

It's tempting to compare the emergence of consciousness in children with the appearance of consciousness in humans, and some of the elements thought to be important are common to both:

language, social life, theory of mind, autobiographical memory and some sort of brain reorganization. But tempting isn't sufficient. We just don't know enough, especially about the history of consciousness in humans. If there's any parallel between the two, it is this: at one point we—as a species and as individuals—weren't conscious, at least in the full modern adult sense, and then we were. How we got there is another thing entirely.

CHAPTER THIRTEEN

Split Brains

"It takes a hemisphere to laugh, it takes a brain to cry"

—Bob Dylan (at least, he should have written this)

WHILE anatomists knew centuries ago that the human brain—as seen from above—appears to be split in half, creating what looks like two identical segments, it wasn't until 1844 that a doctor named Arthur Ladbroke Wigan went beyond the mere observation to argue that where there are two hemispheres, there must be two minds.[1] Wigan put his case bluntly: "The two hemispheres of the brain are really two distinct and entire organs each as complete and as fully perfect for the purposes it is intended to perform as are the two eyes." He bolstered his argument by citing a series of gruesome cases of grievous injury or disease causing near total destruction of one hemisphere but nonetheless leaving the sufferer mentally intact. How could this be unless the remaining uninjured hemisphere was capable of the entirety of human behaviour and intelligence?

Wigan went much further, though, to argue that the two hemispheres could easily house different ideas and wishes, and that while usually one was dominant over the other (the subservient hemisphere contributing as "an assistant aids a workman"), in some cases, when that dominance was weakened, the conflict between the two hemispheres could result in insanity. He even claimed that it was almost always possible to trace, in cases of insanity, two disparate lines of thinking, one belonging to each hemisphere.

Wigan was no idle chronicler: he hoped that when society embraced this idea of two minds in one head that a new attitude would develop, "a merciful feeling towards the great number of unhappy beings who have one brain requiring incessant control; who, with all their efforts, lose their hold from time to time, and commit acts of extravagance and folly."*

Wigan, although his work went unnoticed for decades, could be said to have kicked off the idea that we have two brains in our head, but it wasn't until the 1960s that new and much more dramatic evidence for that claim surfaced. Today, the two-hemisphere brain is central to the consciousness controversy, and the interplay between the two has given rise to some theories of how consciousness might work.

The 1960s was the decade of the first split-brain surgery. Split-brain patients have had the right and left hemispheres of their brains separated by cutting the corpus callosum, the main cable connecting them. This is done in an effort to limit the spread of debilitating epileptic seizures. Cutting the cable confines the seizures to the hemisphere in which they begin and usually makes a patient's life much easier. While the procedure sounds drastic and is considered only after everything else has failed, incredibly, patients who have had the surgery are able to lead virtually normal lives. You would be hard-pressed, if you met one of these people, to detect any outward sign that their brain had been dramatically rearranged.**

---

* The philosopher Roland Puccetti exploited this theme in his book *The Trial of John and Henry Norton*, a weirdly perverse story about a man whose right hemisphere supervised and carried out the rape and murder of his wife; the man's innocent left hemisphere, the hemisphere endowed with language, protested his innocence. You'll have to read the book to find out what happened. R. Puccetti, *The Trial of John and Henry Norton* (London, England: Hutchison & Co., 1973).

**It is true, however, that they often have subtle deficits, like reading difficulties and problems with short-term memory.

It was only when neuroscientists began to study the behaviour of split-brain people in the lab that remarkable properties of their brains were revealed. The corpus callosum, the cable that is severed in this surgery, is hugely important in the normal working of the brain. Something like 200 million nerve fibres cross it, transferring information from one side of the brain to the other—that's more than connect the cerebral hemispheres to the rest of the brain. The irony is that it was only by studying the two hemispheres in isolation, after the cable was cut, that it became clear just how important that cable is.

In an intact brain, information that belongs to one hemisphere or the other doesn't stay that way for long—it is almost instantly communicated to the other via the corpus callosum. But when that cable is cut, the hemispheres can keep private any information that is theirs alone and, more important, can be persuaded to reveal any peculiarities or shortcomings in their abilities.

Thanks to an oddity of brain wiring, pictures, written words and spoken commands can be directed to one hemisphere or another. For instance, a picture that momentarily appears off to your left (as you look straight ahead) will be fed to the right hemisphere through nerve pathways that leave the eyes and cross the brain from left to right. (The reverse is also true: images on the far right are fed to the left hemisphere.) In split-brain patients, those pictures are never shared with the other hemisphere. In fact, two pictures can be flashed, one right after the other, one to each hemisphere, creating a situation where each hemisphere of the same brain possesses different knowledge. Once that happens, bizarre events may follow.

Neuroscientist Michael Gazzaniga has been involved in the split-brain studies since their inception. In one of his most famous experiments, a patient was shown two pictures simultaneously, one positioned to the left, one to the right, so that each was delivered to only one hemisphere. The left hemisphere saw the claw of a chicken, the right a snowy scene. Then the patient was asked to pick, out of a set of additional pictures, one that matched the picture that he had just seen. Of course, asking the patient in this case is a little tricky, because the two halves of the patient's brain have seen different things. But there is a way: ask the patient to point, with either hand, to the appropriate picture. Because the control of hand movements also crosses from one side to the other, the right hand is controlled by the left hemisphere and vice versa.

In this experiment, then, the left hand, responding to the picture seen by the right hemisphere, picked a shovel for the snow; the right hand matched the chicken claw with a chicken. So far so good. But then the stunning step: Gazzaniga asked the patient why he had picked the shovel with his left hand. Asking the patient this question posed a problem for him, because he had been *asked*. Putting the question into words ensured it would be interpreted by the left hemisphere, because the main language areas in this person, as in most of us, were housed there. So his left hemisphere was being asked something it couldn't know: the left hand's choice—the shovel—was made by the *right* hemisphere, and the left hemisphere had absolutely no knowledge of anything the right hemisphere knew. However, rather than saying, "I have no idea," the patient's left hemisphere replied, "You need a shovel to clean out the chicken shed."

A brilliant stroke of invention! The left hemisphere knew only that it had matched a claw with a chicken with the right hand; it had no idea why the left hand had chosen the shovel, but without hesitation it incorporated the shovel into that scenario. This is only one of many such examples of the left hemisphere's ability (and, more important, its desire) to *make sense* of things. Another patient, whose right hemisphere had some limited ability to understand written commands, saw the word "walk" and stood up; when asked why she was doing that, her left hemisphere replied, "I'm thirsty—I was just going to get a Coke."*

---

* There are similar examples from cases of hypnosis. In one, a woman who had been given the posthypnotic suggestion to crawl around on all fours after she woke up, did so, explaining, "I think I lost an earring down here."

Probably the most famous demonstration of the oddities of the split brain is depicted above. In this case the patient was shown two pictures simultaneously. His left hemisphere saw the claw of a chicken; the right a snowy scene. Then he was asked to pick, out of a set of additional pictures, one that matched the picture that he had just seen. His left hand, responding to the picture seen by the right hemisphere, picked a shovel for the snow; the right hand matched the chicken claw with a chicken. But when his left (speech) hemisphere was asked why he had picked the shovel, it had no clue—the left hemisphere doesn't control the left hand. However, rather than admitting defeat, the patient's left hemisphere replied, "You need a shovel to clean out the chicken shed."

There are countless such examples of the left hemisphere's creating a story to explain something it didn't fully understand. Gazzaniga and his colleagues argue that these examples reveal the existence of something in the left hemisphere that he has called the "interpreter," a brain module whose job it is to make sense of—even tell stories about—the bewildering amount of sometimes contradictory information fed to it from a variety of different places in the brain. It's not clear exactly where it is in the left hemisphere (although Gazzaniga suspects the frontal lobe), how much space it takes up or even how it's organized. I'm sure there are neuroscientists who doubt its very existence.

However, if an interpreter or something like it exists, even though it was discovered in split-brain patients, there's no reason we all shouldn't have one. Of course, as long as your corpus callosum is intact you're never faced with such dramatic demonstrations of its operation, but every day, if not every minute of every day, the interpreter is there, doing what it does best: interpreting, rationalizing its way through your life.

What might that involve? Your brain is constantly bombarded with all kinds of sensory and memory data. After the majority of these data are excluded, or processed unconsciously, some of what's left might not hang together; some might be contradictory. But in the end it all must make sense—it must tell a story—for you to be able to react appropriately. In some cases, like the extraordinary ones experienced by the split-brain patients, the interpreter is dead wrong. Most of the time, however, it must be right, or right enough that no one—not you, not anyone around you—notices that it isn't.

Michael Gazzaniga believes that the left-brain interpreter writes the narrative that we call consciousness, but while the split-brain patients are his evidence for that claim, they are also the source of contradiction.

Early on in the split-brain research, the scientists most familiar with it wondered if by splitting the brain they might also have split consciousness in two. It is true that none of the split-brain patients reported that their consciousness had changed as a result of surgery, but then, with expressive language limited to a single hemisphere, you wouldn't have expected them to say anything different. If there were a second consciousness, it would be mute. But that wouldn't mean it couldn't exist (unless you believe, as some experts do, that consciousness absolutely needs language); studies of such patients have at least established that the two hemispheres are very different in style,* and the occasional one has pointed to differences in point of view and even awareness. Those at least suggest that there could be two conscious "minds" housed in one skull.

---

* These differences have been fantastically exaggerated to suggest that you can play tennis better with your right hemisphere, become a better person if you're less left-hemisphere-oriented, and so on and so on. Most of what's written about the right versus left is nonsense, unsupported by any studies.

One of these studies focused on patient P.S., who, because of a brain injury to his left hemisphere early in life, had relocated some language ability to his right hemisphere. P.S. was shown a series of words and asked to rate each on a scale of 1 to 7, where 1 was good and 7 was bad. The list of words included *nice, car, war, kiss, devil, money, mother, sex, vomit, Sunday, truth, hate* and *love*. The striking thing was that the two hemispheres differed in their ratings, with the right hemisphere consistently rating the words higher on the scale, that is, closer to bad. This experiment, and others that have followed, suggests that the two hemispheres can differ in terms of their attitude or outlook, their demeanour. In general, the right hemisphere is more negative than the left.*

Another study revealed that the two sides differ in their approach to problems. Split-brain patients were asked to guess which of two coloured dots would appear on a computer screen. The red dot was programmed to appear 75 per cent of the time, the green dot 25 per cent, but the order of their appearance was random. You can imagine that as the dots begin to appear you're likely to search for the optimum strategy. One strategy would be, because you notice there are more reds than greens, simply to choose the red dot every time. Once you made that decision, you'd spare yourself any further thinking and, in this case, you'd be right 75 per cent of the time. A different strategy might be to try to figure out the underlying pattern of the appearance of the dots. Given that there is no pattern, this is a not a good choice. Here's the funny thing: in these patients, the right hemisphere opted for the simple, straightforward and ultimately winning strategy of going for red all the time, but the left hemisphere persisted in switching back and forth, trying to decipher a pattern that didn't exist. That study is one of several that suggest that the right hemisphere

---

* This allowed some unique interrogation. Once, when P.S.'s left hemisphere was asked what job it would pick, it (he) replied "draftsman"; the right hemisphere, however, spelled out "automobile race." Two different minds in the same brain?

opts for straightforward, uncomplicated solutions to problems, like pigeons do, in contrast to its partner on the other side of the skull.

Gazzaniga blames the interpreter for the left hemisphere's fruitless search for some sort of causal relationship; it can't help but seek a pattern. In fact, in a study where the subjects had suffered strokes to either the right or the left hemisphere, those with intact left hemispheres always opted to search for the pattern they assumed was there, while the one individual with a damaged left hemisphere, forced to rely on the right, adopted the simple and ultimately more successful approach. If your brain is not split, then presumably there is some sort of tension between the two options. As far as I know, there is little evidence to show how that tension is usually resolved, although it is true that humans appear to be the only animals to use the left-brain interpreter strategy of searching for the pattern.

A much more startling display was provided by a young woman whose surgery apparently produced two independent hemispheres that didn't just disagree—they were in violent conflict. Victor Mark worked with the patient during her rehabilitation: "On one occasion she mentioned she did not have feelings in her left hand. When I echoed the statement she said that she was *not* numb and then the torrent of alternating 'yes!' and 'no!' replies ensued, followed by a despairing 'I don't know!'"

When Dr. Mark put a sheet of paper in front of her marked with the words YES and NO, the dialogue between the two hemispheres continued: "When I asked, 'Is your left hand numb?', her left hand jabbed at the word 'NO' and the right hand pointed at YES. The patient became emotionally upset by the lack of unanimity and furiously and repeatedly tried to indicate the 'correct' answer but with the same results. Ultimately the left hand forced aside the right and covered the word YES! The patient by this point was obviously upset by the lack of resolution and I removed the paper."[2] There is a contradiction here. Michael Gazzaniga believes that the left-brain interpreter can explain why, when the brain is composed of a number of separate modules, you have the feeling of one—and only one—consciousness. Information is continuously fed to the interpreter from other parts of the brain, and it is the job of the interpreter to make sense of that flow of information, to create a narrative flow, which, in his words, suggests that "the left hemisphere interpreter may generate a feeling in all of us that we are integrated and unified."[3] However, there is all this evidence of separate consciousnesses in the two hemispheres that has to be explained: if the left-brain interpreter generates consciousness in all of us (as long as our hemispheres are connected), what accounts for the examples we've just seen of right-hemisphere consciousness? There's no evidence for an interpreter there.

No evidence, no agreement. It's clear that the right hemisphere is conscious, but some argue that consciousness there is not as fully developed as that in the left hemisphere. The problem-solving ability of the right hemisphere appears to be much inferior to that of the left, and, more important, it is usually practically speechless. Of all the differences between the hemispheres, this might be the most crucial.

Of course language isn't necessary to have any kind of consciousness at all—what about infants or adults who have become aphasic as the result of a brain injury? (I have communicated, or tried to, with many such adults, and while language has left them painfully incapable of normal conversation, the agony and frustration with which they try makes it absolutely plain that they are conscious.) But researchers like Alain Morin of Mount Royal College in Calgary argue that the importance of language goes beyond conversation: it allows the capacity for "inner speech," the ongoing monologue that we produce silently in our brains.[4] There are testimonials from people who have recovered from a stroke in the left hemisphere that temporarily disrupted their inner speech to the effect that until their inner speech returned, their consciousness was suspended. One, a psychologist named Claude Scott Moss, wrote, "If I had lost the ability to converse with others, I had also lost the ability to engage in self-talk. In other words, I did not have the ability to think about the future—to worry, to anticipate or perceive it—at least not with words. Thus for the first four or five weeks after hospitalization *I simply existed.*"[5]

Moss's account shows that he wasn't literally unconscious in the absence of inner speech, but that he didn't have full consciousness either. There are other such accounts, and they suggest that the right hemisphere, which is normally nonlinguistic, might as a result be conscious, but not fully so. In fact, the case of P.S. lends credence to the importance of speech, because P.S., with linguistic abilities in both hemispheres, is the best example by far of a split-brain person with two separate consciousnesses. Each of his hemispheres seems to have its own sense of self.* The split-brain patients have shown that the right hemisphere experiences the full range of emotions, but, as Morin suggests, it might not be able to reflect on them. Morin likens inner speech to a flashlight in a dark room. You can still find your way around the room by touch, but it's more "vivid and precise" with inner speech.

So if the right hemisphere normally lacks inner speech and is likewise not known to possess an interpreter (assuming the interpreter is crucial), what sort of consciousness might it have? What contribution does that hemisphere make to the whole brain?

The answer is likely to be found somewhere in one of the many scenarios that has been proposed in the past, ranging from the suggestion that split-brain patients still have just one consciousness, to their having two separate but equal consciousnesses, to two half-consciousnesses, to the remarkable idea that we all have two consciousnesses even if our brains aren't split. Alain Morin likes the idea of compromise, that there are two consciousnesses in the two hemispheres but that the

---

* One of the curious things about P.S. is that when his two hemispheres are generally in agreement with each other, he is well behaved and calm, but when they disagree, he is difficult to manage. It sounds a lot like Arthur Wigan's argument that mental difficulties arise when the two hemispheres are in conflict.

one with inner speech, usually the left, has a richer version—deluxe consciousness.*

If this is the case, then your right hemisphere might have a consciousness more like that of a chimpanzee or even your dog. But don't forget, in un-split brains the two hemispheres are in constant communication, so neither hemisphere is on its own as far as consciousness is concerned. Both benefit enormously from the interplay: even if the right hemisphere is nonlinguistic, it still has its own style of reasoning, its own unique way of solving problems. (Much of the 1970s hype about the right hemisphere extolled the virtues of its "holistic" approach to the world.) So the compromised abilities of the right hemisphere that have been revealed in the split-brain patients would not be nearly so obvious in the intact brain. The "junior partner" would still be extremely important.

In fact, there is evidence that while split-brain patients do very well in normal life, away from the probing of psychologists, they are disadvantaged in one intriguing way. Their fantasy life and the exuberance of their dreams are greatly diminished after surgery. If these qualities of the human mental experience are dependent on full communication between the hemispheres, it shows that full consciousness needs not just an intact left hemisphere, powerful as it is, but both. And dreaming, in particular, occupies a central place in consciousness studies, because it is a different kind of consciousness than is routinely experienced. That experience is the subject of the next chapter.

---

* You can't help but wonder what Benjamin Libet's proposed experiment on a volunteer's exposed brain (Chapter 8) would reveal in the case of a split-brain patient. If there actually were a "conscious mental field," you'd think it would envelop the entire brain, but if there are two consciousnesses, one in each hemisphere, there's a problem. What would you have then? Two fields that meet each other at the top of the head? Two overlapping mental fields? It just doesn't work.

Brain Injury and Awareness

It is not just brains that have been surgically split that suggest that the left hemisphere has a special role to play in our mental life. People who have suffered injuries to the right hemisphere, usually by stroke, often have the left side of their body paralyzed (the right brain connects to left side). But in some cases they develop additional, strange conditions in which they deny their paralysis, or even argue that the paralyzed arm or leg does not belong to them. (The technical terms are *anosognosia* and *asomatognosia*.)

Dr. Sandra Black at Sunnybrook and Women's College Health Sciences Centre in Toronto has seen countless cases like this. In one that she related to me a few years ago, a woman in her sixties, a nurse, was admitted to Sunnybrook after complaining of a headache and developing sudden left-sided paralysis. She had suffered a large hemorrhage on the right side of her brain. Four weeks later she was alert with no apparent loss of intelligence. And while her left side was completely paralyzed, she denied that her left arm actually belonged to her:

"Do you have any idea whose arm this may be?"

"Yes, I've looked at the tape [the hospital ID bracelet] on her wrist; she has a similar name to mine. It makes people convinced that it is mine."

"What is the other name?"

"H-Y-N-E-K" [This name has been changed to protect her privacy]

"What is your name?"

"H-Y-N-E-S" (emphasizing the S) They both have the same first name.

"Who is Mrs. Hynek?"

"Before I was put on the ward, there was apparently someone called Mrs. Hynek in my room."

"How do you imagine her arm got onto your body?"

"I don't feel it's really onto my body. It lies on top. It's very heavy, very annoying."

"Did her arm stay behind somehow after she left the hospital?"

"It's the only thing I can suspect. I can't figure out how this happened."

That last sentence captures the poor woman's predicament: she really can't figure out how this happened. She's not in denial—she just doesn't have any information to go on. Michael Gazzaniga sees the left-brain interpreter playing an obvious role in cases like this. Remember that the interpreter gathers information and then tries to make sense of it. If an injury to an arm caused paralysis, the right hemisphere would report to the interpreter that it was lacking sensation from, and was unable to move, that arm, and the left-brain interpreter would incorporate that information, making it clear and unambiguous to the person that the left arm was paralyzed. But when the damage is not to the arm, but to that area of the brain in the right hemisphere that tracks the left side of the body, the situation is dramatically different. The right hemisphere is now incapable of reporting anything to the interpreter, so the interpreter has no knowledge of any damage to the arm. But because it has to maintain the narrative, it resorts to denial of the paralysis.

Neuroscientist V.S. Ramachandran of the University of California at San Diego has also used these tragic cases to gain insight into the left hemisphere's storytelling role, but he puts a different twist on things.[6] Like Gazzaniga, he thinks the left hemisphere is dedicated to weaving together a coherent story from the disparate bits of data entering the brain, but he views the hemisphere as being a little more dogged and indiscriminating. In Ramachandran's mind, it's more of a storyteller than an interpreter, one so dedicated to the story being told that it will incorporate virtually any piece of evidence, no matter how inconsistent. Normally the only thing holding it back is the right hemisphere, which actively monitors the stream of incoming information looking for anomalies, pieces that just don't fit. When one is discovered, the right hemisphere then alerts the left, more often than not persuading it to discard the ongoing story and do a rewrite.

The woman who denied that her paralyzed left arm was actually hers is, to Ramachandran, a case of the left hemisphere's overzealousness. It would rather make up things out of thin air than admit that the ongoing story of two intact limbs is no longer true. Instead it creates bizarre, delusional descriptions that allow it to continue the life story uninterrupted. Indeed, when Mrs. "Hynes" was questioned further, she went so far as to examine the fingernails on her left hand and claim that the pattern of ridges on them was different from those on her "own" fingernails. Even her wedding band seemed somehow unfamiliar. The right hemisphere would, according to Ramachandran, normally interrupt this denial of reality, but the right hemisphere had been so severely damaged that it is incapable of performing its normal regulatory role.

Ramachandran has provided a vivid experimental demonstration of this "devil's advocate" role of the right hemisphere. He used a weird technique that involves flooding a person's ear with cold water. This seems to activate the hemisphere on the other side, prodding it into heightened activity. Ramachandran tried this with one of his patients, B.M., who had suffered the classic left-side paralysis following a stroke in her right hemisphere. At first she denied any disability:

"Can you use your right hand?"
"Yes."
"Can you use your left hand?"
"Yes, I can." (She couldn't.)
"Are both hands equally strong?"
"Yes, of course they're equally strong!"

Her right ear was then irrigated with ice-cold water, but with no apparent effect on her beliefs:

"Can you use both of your arms?"
"Yes I can use both my arms."

But there was a dramatic change when her left ear was flooded with cold water, activating her damaged right hemisphere:

"Can you use your hands?"
"I can use my right arm but not my left arm. I want to move it but it doesn't move."
"Mrs. M. how long has your arm been paralyzed? Did it start now or earlier?"
"It has been paralyzed continuously for several days now. . . ."

If this weren't amazing enough, half an hour later Mrs. M. reverted to her earlier delusional state, insisting that her left arm was normal, forgetting that she had just admitted—or realized—minutes before that it was paralyzed.

Flooding her left ear with cold water had apparently jolted her right hemisphere enough that it had, according to Ramachandran's theory, risen out of its post-injury inactivity to assume its usual role of forcing the left hemisphere to acknowledge her paralysis. Once the cold water effect had worn off, however, her right hemisphere was once more on its own, and delusional again.

Many patients like B.M. eventually recover from their anosognosia and acknowledge their paralysis, but most are incapable of remembering that there was a time when they denied it. One told Ramachandran, "Well, if I did I must have been lying, and I don't usually lie." It sounds crazy, but remember that when you're not aware of something, it's apparently impossible to remember ever having had that awareness.

CHAPTER FOURTEEN

Dreams

I DOUBT there is anyone anywhere who hasn't wondered what it would be like to experience a different kind of consciousness. Legions of meditators and users of "mind-altering" drugs would attest to that. But it's not necessary to expend that kind of effort: just go to sleep, perchance to dream. Dreams are not just mental material crying out for interpretation. They are a second kind of consciousness available to everyone.

We are all hungry for explanations of our dreams—not just to understand the strange or unexpected courses that dreams take, but also for a picture of all dreams: how is it that they can be that curious mixture of very believable and familiar things put together with the entirely unbelievable and alien? And what could possibly be going on in our brains that allows us to accept the weirdness with complete equanimity, only to realize as soon as we're awake how absurd everything was? All of that needs explaining, together with our obvious inability to remember much about any dream unless we record it immediately upon waking.

Why do all these features interest all of us so much? Because each is significantly different from its parallel in waking consciousness. When we're awake (provided we're not in some altered state), each new piece of the scene unfolding before us falls into place more or less logically. Surprising things happen, but they usually make sense after a little explanation. It's rare that we reflect on the events of the previous minutes and say to ourselves, "How could I have just accepted that as something reasonable?" Similarly, it's not usually impossible to remember the events of the day shortly after they happen.

That dreaming *is* a kind of consciousness isn't in doubt: you're aware of who you are (and that you are the same "self" as the one who features so prominently in your waking consciousness) and of what you're doing, and it's also clear that dreaming consciousness draws on waking consciousness for much of its material. People, places and even events are familiar.

Consciousness researchers are very fortunate that dreaming exists. It is the only variant of consciousness that we can experience routinely: dreaming just happens, and it's different enough in these particular ways to make it not just fascinating on its own, but, with any luck, informative about waking consciousness. Can dreaming tell us anything about where in the brain consciousness happens? Are there qualities of dream consciousness that will give us insights into the daily waking version?

The split-brain patients of the last chapter tell us something about parts of the brain that are crucial for dreaming, although in the end it turns out to fall frustratingly short of where we'd like to be. Remember, asking such people about their dreaming is actually asking their left hemispheres what sort of dreams it has, because it is almost always the hemisphere housing the main language centres. Ask any split-brain person a question (unless you take special care to be sure that question is directed to the left ear only) and you're asking that person's left hemisphere.

It turned out that isolated left hemispheres can dream, which apparently came as a surprise to many, because dreams exhibit features that are supposedly best done by the right hemisphere. They are usually dramatically visual, and visual/spatial processing is the right hemisphere's baby. Dreams are also more often emotionally negative, again possibly reflecting the right hemisphere's apparently gloomy view of life. But studies of split-brain patients show that indeed their left hemispheres are still capable of dreaming, with one small caveat: those dreams are dull. In fact, the famous phrase "nasty, brutish and short," although originally applied by the seventeenth-century philosopher Thomas Hobbes to the life of man, is a perfect description of the dreams of split-brain people. Of course, there is really no way of knowing whether the silent companion hemisphere might have been enjoying its own dreams at the same time, or whether those dreams might even have been more like the spectacularly weird ones that most of us have.

But there is a limit to what can be gleaned from studies of split-brain people. They have complicated medical histories: they usually have been on strong anti-seizure drugs (and seizuring despite them) for years prior to their surgery, so it's dicey to assume that their pattern of dreaming, even when divided between hemispheres, is like the rest of the population's. However, in a 1980s review of cases of the dreams of brain-damaged people, Martha Farah and Mark Greenberg found a consistency that mirrored the results from the split-brain patients.[1] Of nine cases over four decades of patients who had completely lost the ability to dream, there was damage to the left hemisphere in eight (and the one case of right-hemisphere damage seemed to be an instance of reversed hemispheric roles). These and other cases indicated that the left hemisphere plays a crucial role in dreaming, although because the vividness of those dreams is much diminished in split-brain patients, it's also clear that the right hemisphere has to be part of the mix to create the kind of dreams most of us experience.*

More recent studies have identified several areas of the brain that play a role in generating dreams; damage to any of them either corrupts dreaming or eliminates it completely. One particularly striking statistic is that around 80 per cent of the hundreds

---

* It hasn't escaped scientists' notice that the idea of a dominant left hemisphere supported by a somewhat subservient right is also the setup for language. The left hemisphere decodes and produces it; the right lends crucial elements of rhythm and tone.

of schizophrenics who were lobotomized in the decades following the 1940s lost the ability to dream. They stand, unhappily, as one of the best pieces of evidence for the importance of the frontal lobes in dreaming. There is still dispute over exactly how many neural centres in the brain participate in, or are crucial to, dreaming, let alone the precise identity of those centres, so at least from this point of view dreaming consciousness doesn't tell us much about waking consciousness.

One dramatic and obvious difference between waking and dreaming is that dreams are not continuous. In fact it used to be thought that dreams occurred only during rapid-eye-movement sleep, periods of which occur sporadically through the night. Despite the name, REM sleep is not just a state in which the eyes are doing something odd; most of the major muscle groups in the body are also paralyzed (ensuring that you can't act out your dreams) and the brain is going crazy electrically. Well, maybe not crazy, but it is as active as it is when it is awake. REM is very different from the rest of the sleep cycle, so-called non-REM sleep. However, in what has amounted to a minor revolution in dream studies, evidence has accumulated over the last two decades that makes it clear that dreams can also happen during non-REM sleep. There is still debate over whether these non-REM dreams are the same, a somewhat simplified version of REM dreams or completely different, and it's still apparently true that the most vivid, complex and memorable dreams occur during REM sleep. Add to this the fact that in some cases of brain damage, patients do have REM sleep but have no dreams, and it's clear that the relationship between REM sleep and dreaming is not as tight and exclusive as once thought.

All of this is complicated by the fact that it's tricky to find out what someone is dreaming and when. In the days when REM sleep and dreaming were practically considered to be one and the same thing, sleep researchers used that connection to gather information about dreams: rather than wait for sleepers to waken in the morning and try to remember a night's worth of dreams, scientists tracked their brain waves with the electroencephalogram and woke them up right out of a period of REM sleep. It turned out to be much easier to remember dreams that way. It does introduce an artefact though: if you wake people from REM sleep and they give you a detailed description of what they were dreaming just before waking, and then you wake them out of non-REM sleep and they are much less descriptive, does that mean the actual dreams were different, or just that it's harder to remember them out of non-REM sleep?

However the relative status of REM versus non-REM dreams is sorted out, research on either can contribute to an understanding of dream consciousness and, hopefully, waking consciousness as well. One curious thing that dream investigations has revealed is that dreams seem to come online much later than waking consciousness does. Until about the age of ten, only about 20 per cent of children report having been dreaming when they are awakened from REM sleep. By age twelve, that has risen dramatically, to about 80 per cent. No one knows what's happening in the brain during that crucial two years that makes dreams change from rare to common.

The content of children's dreams is also unique. Children of kindergarten age, when wakened from REM sleep, tend to report uninteresting, stereotyped, static images of animals, or thoughts about eating. Not only that, their dreams have a much less negative tone than the typical adult dream. These data show that dreaming doesn't achieve full adult form until about twelve, an age at which consciousness has long since been up and running.* It's true that full adult consciousness takes time to establish, but nothing like twelve years. If, as some suggest, the full development of dreams must wait for the brain to finish developing some crucial neural circuitry, why then is waking consciousness not similarly delayed? Why would the dreaming brain require something over and above that which the waking brain works perfectly well with?

---

* Paradoxically, the amount of REM sleep is declining steadily as children develop adult dreams. Infants sleep almost all the time and spend half of that in REM, as opposed to adults, who devote about 25 per cent of eight hours a night to REM. Children hit that adult level at about the age of ten.

Something important must be different about the two states, yet there doesn't seem to be anything about dreams that would require any sort of late tweaking of the brain. There is the weird logic-defying juxtaposition of events in dreams, but if anything that disarray would imply that there's some wiring that's been left undone, not perfected. These child/adult differences seem to me to be one of the most intriguing discoveries about dreams, although the import of them isn't clear.

One very promising approach to understanding dream conscious-ness was based on the chemistry of the brain and how it changes from REM to non-REM sleep. Of course, now that the belief that dreaming is all about REM has been strongly challenged, the pre-eminence of this work has been eroded. But many neuroscientists still buy it, and it could still turn out to be the best lead anyone has to understanding why dreams and waking are so different.

When we slip from non-REM sleep into REM (the first REM period of the four or five that occur in the night usually starts about an hour and a half after falling asleep), there is a wholesale shift in the relative amounts of a number of important neuro-transmitters, the molecules responsible for enabling neurons to propagate their signals to neighbouring neurons. Allan Hobson at Harvard has played a major role in arguing that these chemical changes tell us more about the brain than Freud ever did.*

The chemistry and electricity of REM dreams comes down to this: just before the start of a REM period, the brain is relatively inactive; neurons are firing much more sporadically than during waking. In particular, areas in the frontal lobes responsible for

---

* In case you missed it, here's roughly what Freud said about dreams: the so-called day residue, memories of events that day, disturbs one of the bank of forbidden repressed (usually sexual) wishes lurking in the unconscious. That wish begins to push its way into consciousness, but it is blocked—thanks to the "censor"—and it backs off instead and becomes disguised. Once in disguise, it can pass the censor and enter consciousness as part of a dream, disguised so effectively that we don't recognize the wish at the basis of it.

working memory, attention and reasoning have virtually shut down. The REM sleep episode is kicked off by volleys of signals that originate in the brainstem but spread throughout the brain, reactivating it. In particular, they jolt the thalamus out of its deep slow rhythms and activate emotional areas like the limbic system. The key, according to Hobson and his colleagues, is that this reactivation of REM is different, both chemically and geographically, from waking.

One important difference is that those frontal areas, which were quiescent during deep sleep, are left out of the REM sleep reawakening. They remain virtually inactive, thus robbing the dreaming brain of the contributions that part of the brain makes, which make up an impressive list: self-reflection, time-keeping, abstract thinking, logical decision making and memory retrieval. No wonder there's a loss of reasoning power in dreams! The chemicals that are essential to making memories, such as serotonin and norepinephrine, are greatly decreased, likely explaining why dreams are so hard to remember. Despite these shortcomings, the brain is highly active. It has returned from the depths of dreamless sleep to the kind of activity it has during waking, but it has only the information already in the brain to work with—nothing is coming in from the senses.* So the brain is awake but disabled. Some areas of the brain that aren't normally active at the same time are turned on, circuits involved in vision and emotion are stimulated, and images are formed. Put this together and you have the recipe for the hallucinatory world of REM dreams. Somehow the brain has to try to make sense of it, so stories are made up and the dream is on its way. Allan Hobson and colleagues argue that the changed chemistry makes you believe that you're awake even though you're not, prevents you from recognizing

---

* The fact that our visual consciousness is much less powerful and complete than we think (see Chapter 6) might explain why scenes change so quickly and senselessly in dreams. When we're awake, we can instantly refresh the scene simply by looking at it—so turn away, try to remember what you were looking at, then turn back and it reappears. Unfortunately, the dreaming brain can't do that without visual input. So when you turn your head in a dream then turn back, the scene is likely to be completely different.

how absolutely ridiculous the events in the dream are and also makes it hard for you to remember them when you wake. And it's all being run by the lowly brainstem, whose repeated firing keeps the dream momentum going. As Hobson puts it, "So waking suppresses hallucinosis in favour of thought, and REM sleep releases hallucinosis at the expense of thought." In other words, it's the closest thing to a psychotic episode you hope you ever experience.*

What does all this mean for waking consciousness? That will really depend on whether it turns out that REM dreams are somehow unique, different from the kind of dreaming experienced during non-REM sleep. If not, then the chemical differences don't amount to much. If they are, then one thing this research underlines is the importance of brain chemistry for consciousness. Change the neurotransmitters, and you profoundly change the nature of the experience. Among other things, it makes it hard to maintain the idea of a mind separate from the brain. After all, a separate mind should be indifferent to the kinds of changes going on in the physical realm of the brain. But most neuroscientists don't buy the separation of brain and mind anyway, so for them there has to be a bigger payoff.

Does the chemistry of REM dreaming bring us any closer to understanding consciousness? Hobson and his colleagues think so. They have used their REM investigations to suggest that many forms of consciousness, from coma to waking to dreaming, owe their existence to shifts among three critical factors: activation, input source and modulation. The details are complex, but the outline is clear: as different brain areas become active, our experiences change; if inputs change—that is, if there are

---

* Hobson has written dozens and dozens of papers on the topic of REM sleep and dreaming. The one most pertinent to this idea of dreams and madness is "A Model for Madness?" in *Nature* (1 July 2004). In other writings, he and his colleagues have suggested that "madness" increases through the night: early dreams, both non-REM and REM, being more rational, late dreams being, well, crazy. The irony here is that dream consciousness might shed more light on psychotic states than does normal waking consciousness.

incoming sensory information packets to flavour things, or not—experience also changes; and as the modulation, or chemical control, of neural circuitry varies, brain areas are either reined in or freed to go wild. If you think of these characteristics as three dimensions of consciousness, almost any state you can conceive of—and some that you probably can't—can be described as some combination of highs and lows in all three.

But are dreams useful? This is an important question, related to the even bigger one, is consciousness useful? They're important because if you are in the camp that believes that consciousness is produced in the brain, then you're likely to believe that it is there for a reason: evolution would not have prompted the appearance of consciousness if it had no survival value. But while it has proven difficult to establish a usefulness for consciousness, it has been doubly difficult to apply the evolutionary argument to dreaming.

There are a couple of reasons for this. One is that even if you can prove that REM sleep is useful for consolidating memories of actions and thoughts acquired during the day, is it really necessary for that REM sleep to realize itself as an action-packed dream? That seems dubious: for one thing, there is evidence that skills are learned faster if the people learning them have the time to experience REM sleep in the night or two after learning, but skill learning during waking doesn't require consciousness, so why should skill learning while sleeping? Unconscious learning seems to work pretty well. Also, very little of the dream content that accompanies this memory-reinforcing REM sleep actually has anything to do with the skill being learned.

But there are theories as to why dreaming would have evolved, and one of them connects nicely to dreams that I have had.

In 1996 we bought a family cottage in the woods north of Toronto. The property is more than ninety acres, and beyond that lies uninhabited territory for about as far as you might want to walk or even canoe. The bodies of water are more overgrown beaver ponds than lakes, but it is beautiful country nonetheless.

Several times over the next few years, I had dreams that I would arrive at my cottage and find that a subdivision was being built on my land, or all around my land. There were bulldozers everywhere. These were horrible, depressing, upsetting dreams—adult property-owning nightmares at their best. They were never resolved until I woke up.

I laugh at myself over these, but one theory of why we dream explains exactly why I would have that kind of dream. The theory, proposed by Finnish neuroscientist Antti Revonsuo, argues that dreams are rehearsals of what to do when faced with threatening things.[2] It's easy to forget that for tens or even hundreds of thousands of years our ancestors were hunter-gatherers, exposed daily to threats from other animals and hostile members of their own species. If, Revonsuo says, they conjured up these threatening images in their dreams and then rehearsed appropriate responses to them, they would have been better prepared for life during the day and—and this is important—wouldn't have risked exposure to the actual living threat.*

Revonsuo thinks that the unfolding of the dream story, bizarre though it can be, actually contains too many references to the real world to be nonsensical. The dream world looks a lot like the one we inhabit; we make use of all our senses in the dream; our self is pretty much like the self we know and love—it's all too coherent to be nothing more than a silk purse made out of a neural sow's ear, a

---

* Revonsuo's theory is by no means the only one. Another that I like, but that doesn't seem to be attracting the same kind of attention, was first proposed by Jacques Montangero in 1999. Montangero sees dreaming as a necessary occupation for the thinking brain, which, if it didn't have dreams to chew on, would start on other topics and likely disturb sleep so much that we would continually be waking up. The mind isn't going to sleep, so let it dream! Not content to suggest just one purpose for sleep, Montangero also suggests, not unlike Revonsuo, that dreaming keeps our decision making and planning sharp and ready to function once the day dawns. Nicholas Humphrey, one of the most imaginative of those who think about consciousness, sees a striking resemblance between dreams and play, in that both are rehearsals for real life, simulations of social interactions. But if that were the case, why are there, at least according to Revonsuo, so many animals featured in threatening dreams? Humphrey thinks they're just proxies for threatening humans. J. Montangero, "A More General Evolutionary Hypothesis about Dream Function," *Sleep and Dreams* (1999): 973. N. Humphrey, "Dreaming as Play," *Behavioral and Brain Sciences* (2000): 953.

random bunch of images and thoughts just thrown up by a chaotically reverberating brain.* Dreams to Revonsuo are more like a virtual reality version of waking life.

Revonsuo is a little bit out on a limb with this theory (you always know that's the case when commentators commend someone for having the "courage" to propose what they've proposed) and there are critiques. Some argue that we're biased to remember the most emotionally vivid dreams—such as threatening ones—and that because the majority of the universal human emotions are negative, dreams might simply be reflecting what goes through our minds the rest of the time anyway.

Others have argued that waking sleepers and asking them to report the dream they were having is very different from interrupting TV watchers and having them relate what the program was about. We have all had the experience of having a vivid dream disintegrate in the seconds after waking, and it's not unrealistic to suggest that in the struggle of the still drowsy and disorganized mind to remember, key facts will be forgotten, but more important, given the brain's capacity to make up a story, facts not actually dreamed could be incorporated into the telling of the dream story.

Related to that idea is the argument that anything that is threatening in a dream is more likely to be remembered because it is more vivid and salient, although Revonsuo counters by saying that waking memory is prone to the same biases and yet doesn't contain the same frequency of threatening ideas.

---

* However, philosopher Eric Schwitzgebel has argued that we can't be as sure about the sensory qualities of our dreams as we'd like to think. His review of the dreaming literature suggests that in the first half of the twentieth century, most people seemed to believe that they dreamed in black and white, with the acceptance of dreams as coloured being a much more recent phenomenon. He attributes this to the fact that media (photography, movies and television) were largely black and white until the 1960s, and that people unwittingly took those media as models for their dreams. In fact, he wonders if our dreams have any colour at all, in the same way that objects in a novel, unless explicitly stated, are not "coloured" by us as we read about them. And as something to think about, Schwitzgebel wonders if when we finally have media that give us touch sensations of what we're watching, we'll start to realize that we have touch in our dreams. Eric Schwitzgebel, "Why Did We Think We Dreamed in Black and White?" *Studies in History and Philosophy of Science* 33 (2002): 649–60.

So in this light, my cottage dreams make perfect sense—to a point. I did dream about threats that would ruin exactly what was important to me about the cottage—the land around it—and it was a sneak attack, something that happened without my knowing it because I wasn't there all the time. All things to fear, but there are a couple of inconsistencies with the theory. I didn't develop much of a response in my dreams; most of them simply left me completely frustrated and unhappy. Another oddity is that these were dreams about very modern concerns: ancient hunting-gathering hominids never worried about rapacious developers or bulldozers. However, there are answers to my concerns. First, the fact that I wasn't able, at least in my dreams, to come up with a response doesn't necessarily put the theory in a bad light, although critics have pointed out that the common experience of being paralyzed as the train bears down on you is hardly a rehearsal of a response you'd like to put into practice. But Revonsuo argues that in analyses of dreams he's done, more than 90 per cent of the responses are appropriate, very few being irrelevant or impossible.

As far as dreaming about things modern, Revonsuo sees the threat response mechanism working on anything that is vivid in one's memory or thought processes, and given that most of us aren't fleeing from attacks by carnivores but are instead worrying about the threats in modern society, it makes sense that I would dream about that. Then why would we dream at all about the threats that faced us long ago? Those dreams would be the "default" content, to be put into play when there's nothing more threatening for the dream system to chew on.

That's one approach to dream consciousness: to believe that it exists for a good reason. On the other hand, it's possible to take the polar opposite view, and ask why dreams should have any purpose at all. Those like Revonsuo who believe that dreams were selected by evolution have to acknowledge that possibility, that all that mental activity at night is just some sort of incidental spinoff from what's going on in the brain anyway.

Philosopher Owen Flanagan is one of those dream skeptics.[3]

To him, dreams are just the "noise the [brain] system creates while it is doing what it was designed to do." He alludes to the brain's capacity for creating sensible narratives by arguing that while the individual dream moments are that noise (he calls them "chaotic neuronal cascades"), they are put together by the brain into some semblance of a story. Flanagan attributes this to the wonderfully creative storytelling ability of the cerebral cortex (reminiscent of both Michael Gazzaniga's and V.S. Ramachandran's descriptions of the roles of the two hemispheres) working with meaningless material. In Flanagan's view, the brain does that with dream content just because it does that with *any* neural activity going on in it. So in this view, your dreams are meaningless, at least until you give them some meaning. Once you say, "I dreamed that because . . ." you're putting into the dream the only meaning it ever had.

Flanagan is careful to differentiate between rapid-eye-movement sleep and dreams. He accepts that REM sleep has some useful functions that would have been selected by evolution—the overnight restoration of depleted levels of brain neurotransmitters, the deletion of unnecessary information. But in his mind, neither requires that we dream while it's happening. As he puts it, while it might be necessary to cycle through REM sleep to learn a set of nonsense syllables, there are no reports of subjects in such experiments actually dreaming about those nonsense syllables.

He concedes that it is possible that dreams have some use, even if those uses didn't evolve under natural selection. The elements of the dream are noise, but the way we put the dream story together to make sense of that noise represents a uniquely personal view of our own brain at work. Flanagan suspects that if your brain and mine were given the same random dream elements, we would create quite different stories to accommodate them, and those stories, stylistically our own, could be very useful to both of us—they would give us a glimpse of our own brains at work, and might even lead to our thinking differently about ourselves. If that were the case, then dreams would indeed have a use—or, more correctly, would have *turned out* to have a use—but it wouldn't be a usefulness that prompted the evolutionary selection of dreams in the first place.

The problem I have with this idea is that I don't think it explains why I would dream of subdivisions on my cottage property, or why, at least for a while, those were the only unpleasant dreams I was having. The elements of those dreams, while different each time, were nonetheless so closely related that it would have been impossible to argue that each dream was my personal assembly of unrelated items.

The debate winds back and forth, and it's not clear at the moment how it will turn out. However, Revonsuo uses one of Flanagan's own dreams to strengthen his case. Apparently Flanagan dreamed, when he was five, of being chased by a pack of wolves. Flanagan attributes the story to a dreamy reminiscence of the tale of the three little pigs, but Revonsuo doesn't buy that, because the wolves are so different: the ones in fairy tales are more like bloodthirsty intelligent humans who happen to be wolves. Revonsuo argues that if the threatening animals in our dreams all came from fairy tales, our dreams would be populated by wolves that "huff and puff and dress up as grandmothers."

Revonsuo's theory is just one of many on the role of dreams, and it runs parallel to the argument over the purposefulness of consciousness in general. But so far dreams haven't shed much light on waking consciousness, probably because dream consciousness itself has not been well explained. The key to connecting the two must surely lie in the peculiarities of dreams: the combination of familiar and bizarre, the unquestioning acceptance of the dream events, no matter how odd, and the fragile dream memory. In those ways, dreams are different from waking and even from daydreaming, which, although apparently free to move randomly from subject to subject, never connects the dots in the same way dreams do.

Finally, the picture is complicated by the ability of some people to have lucid dreams, dreams in which they are able to realize that they are dreaming and can signal that awareness to dream researchers by, for example, a set of predetermined eye movements. Lucid dreaming represents an intrusion of waking consciousness into the dream but without actual awakening. Some experienced lucid dreamers can even control the course of their dreams. While most of the lucid-dreaming literature stresses the otherworldly, cosmically enlightening aspect of the phenomenon, lucid dreaming is fascinating from the consciousness point of view precisely because it represents something partway between waking and dreaming. And because of that it is hard to understand.

For instance, if you agree with Allan Hobson's view that the chaotic craziness of dreams results from the chemical changes of REM sleep together with inactivation of the frontal lobes, then lucid dreaming must somehow restore some of the waking state, or lucid dreamers would not be able to make decisions or carry out plans in their dreams as they apparently routinely do.

Also, if all of the eye movements of REM sleep are randomly generated by impulses coming from the brainstem, how is it possible for lucid dreamers to move their eyes to signal that they are now dreaming lucidly? Something has happened to inject some qualities of waking consciousness into the dream, and while there have been experimental demonstrations that lucidity is a real phenomenon that occurs during REM sleep, and apparently only in REM sleep, there still isn't any clear evidence as to how—or where—it differs from regular dreaming. Gathering such evidence will be difficult: much of the evidence for chemical changes in REM sleep has come from animals, which can't testify that they are having lucid dreams, and there's no guarantee that brain imaging in humans would reveal the differences between lucid and normal dreams. However, such differences, if discovered, would be precious data to have, because they would represent differences between waking and dreaming consciousness. Freud would be fascinated to see that.

aside, you can imagine how tempting these technologies are
can talk all you want about how consciousness feels or wha
like to taste a fresh peach, but if you can see what's going on in t
brain *as that happens*, then aren't you right there at the threshol
of explaining it all?

Well, not exactly. While this is a great and tantalizing prospect, it
is not as straightforward as it sounds. The problem is that after
centuries of questioning and investigation, no one yet knows
exactly what consciousness is or, to be more specific, what might
be happening in the brain (and where) to produce it, and that
makes it much harder to search for it. You don't want to be in the
position of the drunk searching for his car keys under the street-
light not because he lost them there, but because the light is better.

Antti Revonsuo, the man who suggested that dreams are
rehearsals of threats, has voiced doubts about the ability of imaging
technologies to discover consciousness.[1] He questions whether the
tools at the neuroscientist's disposal are up to the task of identifying
something the nature of which is so uncertain. Revonsuo points out
that there are differences of opinion about where to look. If I toss
you an apple and ask you to catch it then take a good look at it,
where in the brain are you conscious of that apple's appearance?
There is the fact that the different aspects of the apple—its colour,
shape and the arc it described as it flew through the air—are all
resolved in different places in your brain; some would argue that
nevertheless, somewhere, somehow, the overall "appleness" of it
would be represented. Some would even say there *is* no particular
place. So where to look is a puzzle.

There is also the question of time. There is a suspicion that
consciousness might involve not just neurons firing, but neurons
firing in many parts of the brain simultaneously—not just at the
same time, but in synchrony. There's an issue of rapidity as well.
We are capable of switching attention on millisecond scales.
Revonsuo suggests a demonstration that provides yet another
excuse to sit down in front of the TV. Turn the TV sound off, close
your eyes and start changing the channels. As each new channel

CHAPTER FIFTEEN

Images

W E LIVE in an era that would have been unimaginable for the pioneers of the search for consciousness, a time when we can actually produce images of the brain, pictures not just of its physical structure, but of neurons at work. You'd think armed with that ability it would be a simple matter to take pictures of brains that are experiencing consciousness and compare those brains with ones that are not. The differences would represent the first images ever of consciousness. William James would be entranced; Descartes, stunned.

There are several technologies available to reveal the workings of the brain; those most often recruited for consciousness studies are positron emission tomography (PET), functional magnetic resonance imaging (fMRI) and electroencephalography (EEG).* The choice of technology is driven at least partly by the goal: if you're after the precise timing of an event in the brain, EEG is best; if location is what you want, PET and fMRI are the choices. The converse is true as well: PET and fMRI are not good at timing, and EEG can't pin down location very well. Those weaknesses

---

* There is also the related technology called event-related potential, or ERP, which links a specific part of the complicated and confusing welter of activity recorded in an EEG to an event, such as the presentation of an image or a spoken word.

clicks in, open and close your eyes as quickly as you can. You have probably never seen the images before, you don't know what they're going to be, and yet you can easily identify every one. A tenth of a second is usually enough, meaning that consciousness must be able to change its focus that rapidly. So it's dynamic and hard to localize, an elusive target indeed.

These are substantial technical demands. Revonsuo cautions that as far as he is concerned, the current technologies are far from being able to identify brain activity that actually constitutes consciousness. At best, they can identify brain activities that are somehow involved in or *related* to consciousness, not those that represent the final answer. Although they do represent a dramatic upgrade on the wishy-washy statement that consciousness is created by the brain, they will, in Revonsuo's opinion, fall short of a precise identification of the mechanisms underlying consciousness.

Take EEG recordings, for example. These have the virtue of being extremely time-sensitive: they can record electrical events lasting only a millisecond, a thousandth of a second. But there is a certain vagueness about the electrical patterns produced by the EEG. For one thing, they are spatially smeared out because they have to travel from the brain through the skull to the recording electrodes. You can show that a particular wave pattern started in a roughly defined area, but you can't trace it back exactly to the cluster of neurons that generated it. Revonsuo sums it up by saying, "what we are seeing are gross spatial averages of fast changes in the synchronous activity of millions of synapses whose location we cannot observe." So while the EEG has been used to reveal waves of electricity passing through the brain that quite likely have something to do with consciousness, at this point it is practically impossible to go much further than that (although see Chapter 16).

The frustration of EEG is that you can't precisely identify the whereabouts of the source of all that electricity. But what about PET and fMRI? They are all about location. They don't actually pick up on the electrical discharges of neurons, but instead record a proxy of their activity: the increase of blood flow in a particular area of the brain. The picture is actually more complicated than that, but scientists accept that when a part of the brain "lights up," the neurons in that area have suddenly become active. "In that area" is deliberately vague: the blotches of colour revealed in these brain images can represent an area of brain tissue no bigger than a millimetre square, but that is still enough space to contain millions of neurons, with tens of millions of synapses. That's one limitation. The other, as I indicated earlier, is time. It takes a second or so to create an image of brain activity, but the activity itself is operating on scales hundreds of times shorter. The images produced are analogous to photos, whose long exposure times make it impossible to capture movement.

Revonsuo has the same sort of complaint about these imaging techniques as he does about EEG: they are showing us something, but that something falls far short of an image of consciousness, and it may be that the technologies are simply inappropriate. If consciousness somehow involves neurons in far-flung areas of the brain adjusting their firing rates to synchronize with each other, it's hard to see how that could be captured by fMRI.

Revonsuo's critics (there is never a discussion of consciousness without critics) complain that he is taking the "glass half empty" point of view, ignoring advances that have been made in the technologies he dismisses, and also that no one who's familiar with imaging technologies would ever make the claim that they'll provide the final answer anyway. Regardless, in case you thought that brain imaging was going to show us exactly what consciousness is, you're likely to be disappointed. But it can reveal some of the things that are happening in the brain *when* we are conscious (the now-famous phrase, "the neural correlates of consciousness"), so, with the appropriate caution in mind, let's see what brain imaging has shown us.

One of my favourite experiments—one that is really, really clever—uses a famous illusion, Rubin's vase, to track down one aspect of consciousness. Rubin's vase is a particular good illusion to use, because it's an ambiguous figure (like the Necker cube of

Rubin's vase is another ambiguous figure like the Necker cube, but better for consciousness testing because the alternatives—a vase or two faces—are perceived in different parts of the brain. Clever experiments and good brain imaging might reveal which places are active when one or the other is dominant.

Chapter 6). As you look at Rubin's vase, it switches from being a vase to two faces, in profile, looking at each other. It's a beautiful coincidence that the human brain processes faces (which are of utmost importance to a social creature) in a different place than it does objects. Rubin's vase also has the advantage that, like other ambiguous figures, it switches from one version to the other on its own, without any input from us. What happens in the brain when that switching occurs? A team of British scientists led by Colin Blakemore found out.[2]

They put volunteers into an MRI machine and then showed them the Rubin's vase illusion. As you'd expect, these people had the experience of the illusion switching from one form to the other. Whenever that happened, they pressed a button. The exper-imenters were then able to correlate the button presses—which indicated which version of the illusion the subjects were experi-encing—with the images provided by the MRI. The results were partly what you'd expect. Each time volunteers reported seeing the vase switch to the faces, an area called the "fusiform face area" lit up. This is an area that in humans is apparently specialized for the perception of faces. The reverse, however, wasn't true: there was no discernible increase in the area known to be sensitive to inanimate objects when the illusion changed from face to vase. The experi-menters suggest that even though they had demonstrated that the object area does light up when volunteers viewed unambiguous objects, such as lamps and microscopes, it might be true that the vase version of the illusion fails to provide enough detail to excite this area. Even the face in the illusion didn't excite a second face-sensitive area as much as did single photos of faces, possibly because that area is sensitive to details of the face—direction of gaze, lip movement and expression—none of which are present in the illusion.

This is a lovely experiment, built on the fortuitous fact that this illusion, unlike most others, activates different parts of the brain. Necker's cube wouldn't work, because a cube that sticks out one way isn't going to activate an obviously different part of the brain than a cube that sticks out the other way. But what exactly does this neat experiment tell us? It shows that a well-defined area of the brain ramps up its activity when people are aware of the faces in the image but not when they are aware of the vase. The suggestion that there might not have been enough detail to fire up the "vase" area, if true, suggests that in some circumstances confusion would reign if the alternative versions were very similar. I'm sure I experienced the subjective side of that just before I wrote this chapter.

I was driving along a country road near my cottage north of Toronto, in winter, when I noticed a dog standing by the side of the road ahead. You have to be careful on roads like this because the local dogs feel it's their right to cross at any time. Or *was* it a dog? Closer inspection revealed that it had to be two mailboxes, one behind the other, situated so that the two posts stood slightly apart, like legs, and the two boxes themselves overlapped slightly so as to appear to be a body. But they looked very doglike. The funny thing was that the ambiguity didn't resolve itself as I got closer, they didn't become definitively mailboxy, until, when I was almost on top of them, the mailboxes . . . turned out to be a dog after all. It's unlikely anyone has done an MRI of people looking at mailboxes and dogs, but I'd be pretty confident that those two objects would be processed in different parts of the brain, and I can only imagine the confusion in my temporal lobes as that incident unfolded.

So the activity in those areas of the brain have something to do with consciousness, but remembering Revonsuo's cautions, it would be too daring to say exactly what that meant. You should be careful to resist the temptation to say that activity in the fusiform face area is all that you need to become conscious that you're looking at a face, but clearly it plays a central role.

There are other imaging experiments that reveal activity in isolated parts of the brain that correlate with awareness. The area near the back of the brain called MT is responsible for the perception of motion. (You might remember from Chapter 3 the story of the woman who suffered damage to that area and was evermore incapable of seeing things move.) MT is crucial to something called the waterfall illusion: if you stare at something, like a waterfall, that's moving continuously in one direction, then switch your gaze to a stationary scene, that scene will appear to move in the opposite direction. It's a compelling illusion, and brain imaging has shown that when people experience it, MT is active. It's active even though the person is looking at something stationary. When the activity ceases, as it does after a few seconds, the experience of movement disappears. Is MT all that's needed to perceive motion?

Nancy Kanwisher of MIT suggests that you imagine the following experiment.[3] It is easy to change a monkey's perception of motion by stimulating a small area of MT in the monkey's brain. That being the case, how about removing a chunk of MT—even from a human brain—being careful to keep it alive and intact, then stimulating it while it's sitting there in a dish on the lab bench. Will that detached piece of brain "experience" motion? How exactly? It just doesn't make sense: for consciousness to be the rapidly shifting interconnected thing that it is, areas like the fusiform face area and MT must connect to far-flung parts of the brain.

However, what it clear from these studies is that those isolated areas do play an important role in conscious experience. But there's more to it than that. Marcus Raichle of Washington University in St. Louis and several colleagues have spent years imaging the thinking brain, and they have shown that even very simple tasks have something different to say about consciousness.[4]

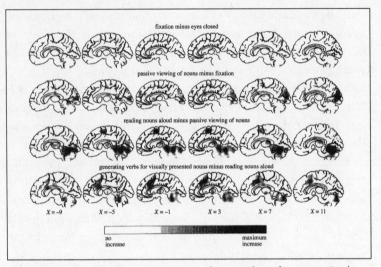

fixation minus eyes closed

passive viewing of nouns minus fixation

reading nouns aloud minus passive viewing of nouns

generating verbs for visually presented nouns minus reading nouns aloud

X = −9   X = −5   X = −1   X = 3   X = 7   X = 11

no increase   maximum increase

These images reveal the brain activity seen when people perform even simple tasks. Darkening indicates activity that was not present in the previous task.
a) The appearance of single nouns on the screen (second row of images) triggered a welter of activity in the visual areas of the brain, even though participants had simply been told to maintain their gaze on the screen. It's clear that even a single word triggers multiple reactions.
b) Different activity was produced when the words were spoken aloud (third row), some being those areas responsible for the organization of speech and the movements of lips and tongue to produce it.
c) But the most interesting images were produced when the people in the scanner had to think of a verb to go along with the noun (fourth row). There was suddenly new activity all over, some of it in areas far removed from the visual cortex at the back of the brain.

They looked at what goes on in the brain when people are presented with single words—nouns—on a screen in front of them.
The questions the experimenters asked were, what differences
are there between simply seeing a noun, speaking it and coming
up with a verb appropriate to that noun? I wasn't kidding when I
described these as simple tasks: imagine how much more complicated it would be to read a newspaper article aloud and then
discuss it with someone.

However, simple as they are, these tasks establish some important facts about consciousness. First, even the appearance of single nouns on the screen triggered a welter of activity in the visual areas of the participants' brains, even though participants had simply been told to maintain their gaze on the screen. It was unclear exactly what those several areas of brain were doing, but obviously a single word triggers multiple reactions.

More activity was produced when the words were spoken aloud, and some of those areas could be identified as brain regions responsible for the organization of speech and for the movements of lips and tongue associated with it. But the most interesting images were produced when the people in the scanner had to think of a verb to go along with the noun. There was suddenly new activity all over, some of it in areas far removed from the visual cortex at the back of the brain. Remember, the only difference between this and saying the noun out loud is that producing a verb requires thinking and innovation, as opposed to the well-practised art of reading aloud.

Again, no particular surprise: when something out of the ordinary is required, you have to recruit more brainpower to deal with it. But this became even more interesting when the participants were allowed to practise their verb-making. As they became more skilled, and even stereotyped (by saying the same verb in response to a repetition of a noun), those areas that had lit up frantically when the task was novel began to recede into the background activity, and areas that had been prominent in the simple, almost reflexive routine of saying whatever noun appeared on the screen (but then had faded during the verb exercise) re-established themselves.

Was this a demonstration of the idea that consciousness comes online when we are faced with something unfamiliar but then moves on as soon as we become accomplished at it? It appeared to be the case, and a parallel experiment seems to confirm it. In this case, people had to learn to move a pen around a maze blindfolded. (The maze was delineated by raised edges.) In the initial learning phase, the activity of some areas was dampened while other areas were stimulated, but, as in the noun experiment, once the maze was mastered, the pattern reversed. The actual brain areas were different in the two studies, so while the principle holds that consciousness steps in when learning is required, the identity of the areas involved is likely different for every different kind of task. Yes, in some ways consciousness is predictable, but in others, it's even more complex than we might have feared.

One final note from the Raichle studies: even when people were lying in the scanner with their eyes closed or staring blankly at a dot in the middle of a TV screen, there were areas along the midline of the brain that were highly active—not just humming along like a motor idling, but burning glucose and consuming oxygen well above the resting state. These are areas that decline in activity once the brain meets the challenge of the task but resume once practice has made perfect. It's likely that these are areas that monitor your surroundings on a continuing basis, processing sensory information, on the lookout for anything meaningful. As soon as the brain has to focus its attention, as in thinking of verbs to go with nouns, those areas step aside, only to come back when you're able to accomplish the task at hand without thinking much about it. There is, I think you'll agree, an interesting parallel here with what you'd experience. You'd start being conscious of what's going on around you, suspend that line of thought to concentrate on the task and eventually, once the task is well in hand, shift back to your normal musings. There would be no break in your consciousness, just changes in focus, changes that can be seen in the images.

Other scientists took this evidence that there are brain areas that stay active even when you're not really attending to anything in particular and wondered if those areas might have something to do with that key feature of consciousness, the maintenance of a "self." When you pause to think, you know it's *you* that's thinking, and that the same you has existed in the past and with any luck will continue to do so in the future. If all aspects of consciousness are generated in the brain, then somewhere some neurons must be engaged in keeping that concept of self alive and well, and it would be reasonable that such neurons would be continuously active, at least when nothing else is preoccupying the brain.

A Danish team took an unusual approach to an investigation of this idea.[5] They asked volunteers to think about their own personality and physical appearance contrasted to the personality and appearance of the queen of Denmark, so chosen because it was unlikely that any of the experimental subjects would have had anything to do with her personally but would at the same time be familiar with her. When the volunteers were in the scanner, they were simply to think, but afterwards they explained what they had been thinking about. Here are some examples of those thoughts:

"I am extrovert and talkative, while on the other hand I have sides only my friends know. I am rather loving to my girlfriends and my family and that like. And considerate. On the other hand, I can be quite cold sometimes."

"She seems open and forthcoming. She seems like a person with authority—a good representative of Denmark. She appears to be a queen we can be proud of. She is creative and interested in art and in meeting people."

"I have dark hair and brown eyes. And a quite large nose and big ears. I thought about my lips and how my body is shaped. . . . Er . . . and then there is my belly . . . I have relatively large hands."

"I thought she must be long-haired even though you never really see it. The hair is always tied in a knot. She has a rather long neck, and a long face, and normal ears."

The PET scans of the subjects' brains when they were thinking the above thoughts revealed a consistent difference between self-thought and thoughts about the queen. Some of the areas involved in thinking about self were close to or even identical to the areas that Marcus Raichle and others have found to be active when nothing significant is going on in the brain. The authors of the study argue that the maintenance of a sense of self represents a cooperative activity of sensory parts of the brain towards the back and "executive" parts at the front; in their words, self represents a "summit" of these two areas.

How strong is this evidence that these areas are really the seat of the self? This research contradicts other work that points to the right frontal lobe, not to these areas running down the middle from front to back. Those experiments involved scanning people who were looking at pictures both of their own face and the faces of others, or at novel faces made by morphing their own faces with others, and found that the recognition of one's own face was almost always associated with greater activity in the right hemisphere. This research follows on the heels of the mirror-recognition tests (see Chapter 10), which have been used to argue that any creature that realizes that the image in the mirror is its own reflection is conscious, or at least self-conscious.

Contradictions such as these have the effect of curbing your enthusiasm for any one explanation, and that's probably a good thing. Without repeating Antti Revonsuo's warnings about imaging and consciousness, it's worth emphasizing that while brain images might be a window into the brain, they are a small, dingy window looking into a not very well lit room. While I haven't referred to any one experiment that uses EEG or ERPs (event-related potentials, an electric technique with fantastic temporal resolution) to identify conscious events in the brain, you can be sure that what they do show, with their far superior ability to track rapid events, is that any single mental activity is accompanied not only by activity in several areas of the brain, but by activity that rockets around the brain at breakneck speed. We haven't yet seen the ideal imaging technique, one that would combine the best attributes of all of them and bring us the first, really detailed, temporally fine images of what goes on. We might have to wait for those to resolve these arguments like the one over where the self resides in the brain.

Two final thoughts: it's a good idea not to forget that though tracking milliseconds of brain activity over tiny areas of tissue gives the impression of fabulous precision—the electron microscopy of thought—contained within that tiny time frame and miniscule area are thousands if not millions of neurons, each one of which makes hundreds, or even thousands, of connections to other neurons. How easy it would be to miss something significant! The other factor is chemistry. Neurons communicate by releasing neurotransmitter molecules. There are dozens of different kinds; many have multiple actions; their amounts change drastically in the transition from sleep to dreaming to waking; they are known to play key roles in the reaction to drugs and the development of a variety of mental illnesses; but there is very little that has been said so far about the chemistry of waking consciousness. That too will have to become part of the picture.

CHAPTER SIXTEEN

How It Works

A S FAR as I can tell, there is only one thing that consciousness researchers and commentators agree on: there is still much we don't know about consciousness. Some would even say there are things we will never know. But of course, as is true in any kind of science, having only a partial picture hasn't prevented a proliferation of theories about how consciousness works. That is, after all, what theories are all about.

However, enumerating the theories that exist would create nothing more than a catalogue of ideas, without any sense of which ones have caught the imagination of researchers, which have sunk like a stone and which are openly derided. In this chapter I'll focus on one scenario, called the global workspace, and try to show why it currently is one of the most popular approaches to an explanation of how consciousness works.

First, what does any theory of consciousness have to do? It should establish clearly where the line is drawn between unconscious and conscious mental activity; it should map the parts of the brain whose activity correlates with consciousness; it should fit or make sense of attributes of consciousness such as its highly selective nature, its abandonment of mental activities that have become routine, its alteration by brain damage or surgery and, ideally, the most difficult challenge of all—the quality of experience, the qualia associated with consciousness. However, there are few—if any—theories that explain qualia, at least to the satisfaction of the consciousness community at large. There are some ideas about why we have the vivid experiences of things we perceive, like the colour of a rose or the taste of an orange, but no idea how they actually come about, so that requirement should be struck off the list if we're going to be realistic at all.

These, then, are some of the known features of consciousness that should be addressed by any theory of how it works. At the same time, assuming that consciousness is useful to the organism with it, a model of consciousness should provide for the following: processing of information from the senses, melding that information together and using it to create knowledge of the surroundings that allows for action and reaction to events and, at least in higher organisms, the ability to use that information together with memory to plan, choose among alternate courses of action and interact socially. That is one huge menu of requirements, but even if a theory is unable to provide for all of them, it at least can't be inconsistent with them.

Global workspace theory began in work in artificial intelligence. It describes a way of organizing artificial systems to make them as efficient as possible. While neuroscientist Bernard Baars of the Neurosciences Institute in San Diego was probably the first to enthusiastically apply these ideas to consciousness and is certainly associated with this idea more than any other, there are now several researchers who either have published evidence that supports the global workspace or have acknowledged that it represents an advance, if not yet a complete theory. Even Daniel Dennett, who, fairly or unfairly, I seem to have characterized in this book as a doubter, has acknowledged that "theorists are converging from quite different quarters on a version of the global neuronal workspace model of consciousness."[1]

So what is the global workspace? Bernard Baars has used the analogy of a theatre to explain it, fully aware that any sort of theatre reference in consciousness runs the risk of being identified with the disgraced idea of the "Cartesian" theatre I mentioned in Chapter 1.[2] But this is not a theatre in which all of consciousness is presented for the viewing pleasure of a single individual, a homunculus, your "self." In the global workspace theatre, the spotlight on the stage highlights the material that is currently in consciousness.

That spotlight is driven by attention, as you can easily convince yourself by the old trick of diverting it from reading these words to other elements of your environment that, up until now, you have been ignoring, such as background sounds, the pressure of the chair against your body, the faint tastes in your mouth. Shift your attention back to the words, and those things that briefly entered your consciousness now recede into the background again. It has been suggested that the prefrontal areas of the brain control this attentional spotlight, but this control can be overridden by alerts from other parts of the brain that are monitoring the environment.

The material that is spotlighted on the stage is broadcast, or made available to, the audience, which is dark and unconscious and, as in most theatres, occupies a much larger space than the stage. In addition, the material on stage is controlled or shaped by what is happening backstage. Baars emphasizes that, again as in a real theatre, what is on stage at any moment is a distillation (and a short-lived one) of information from many different sources and, similarly, at every moment what's on stage flows out to the audience for them to deal with in whatever way they can.*

Translating that theatrical image into neuroscience yields the following picture of consciousness: whatever occupies the stage at any moment is what's in your consciousness *at that moment*. It is your working memory. Working memory is the

---

* If you're wondering why it should be a good thing to be aware of only a fragment of the mental activity going on in your brain at any time, there is a good analogy from the world of DVDs. Is it really better to experience all the outtakes, random interviews and useless background that comes with the DVD, or to enjoy the original movie as it was? Now extend that to the idea of a DVD that includes a detailed explanation of all the camera mechanisms, the scheduling of meals, all the e-mails that were sent back and forth before and during filming and the financial records, and you'll have to agree, less can be more.

same thing as short-term memory. It has different parts: an inner auditory component, where you keep alive in memory verbal information you're working with, such as the phone number that you mumble to yourself as you make your way to the phone; a "visuo-spatial scratchpad," where images are momentarily preserved; and some kind of executive to run the whole thing, which might lurk backstage. As you read this, you are using working memory to keep the ideas of this sentence and the last few "in your mind" at the same time: working memory needs you to be conscious. There is no evidence that working memory can run unconsciously, and a moment's reflection (which of course requires working memory) makes that obvious. How could you keep track of what you're talking or thinking about if you weren't conscious of it?

Working memory may be relatively short-lived, but not necessarily: you can keep that phone number in consciousness for some time if you keep refreshing or rehearsing it. On the other hand, there is a minimum. If a piece of information is not attended to long enough, it will fail to reach consciousness, the equivalent of edging onto the stage but having the spotlight miss you, thus precluding the momentary opportunity for conscious fame. This means there are actually two requirements for a stimulus—say, an odour—to reach consciousness. First, it has to last long enough to be detected by the neurons in your nose, and second, that resulting activity has to last long enough to reach consciousness.

What's on the stage represents, as we already know, a trickle of the total amount of information in your brain that is eligible for selection. The "audience" is a huge array of countless unconscious modules, separate containers or circuits of information in the brain of which we are unaware. Among these are modules responsible for some things that never make it into our consciousness, such as the rules of grammar that allow us to understand sentences, like this one, which we have never seen before, or to begin the processing of hundreds of visual images, most of which slip by unnoticed.* But there are also modules that hold information that we can bring into consciousness—the audience contains them all. These unconscious elements both contribute information to the stage and are recipients of the information that is presented. Global workspace theory emphasizes this last point: the job of consciousness is to broadcast its contents to the rest of the brain.

Baars once described global workspace architectures as "publicity systems in the society of specialists. Input specialists (such as perceptual systems) compete for access to the global workspace; once they gain access, they are able to disseminate a message to the system as a whole. When a problem has a predictable solution, it is handled by a specialized processor. But when novel problems arise, or when new, coherent actions must be planned and carried out, the global workspace becomes necessary."[3]

The theatre image, although an appropriate metaphor for a publication like the *Journal of Consciousness Studies*, doesn't convey the disparity in size between stage and audience nearly as well as an outdoor rock concert like Woodstock or, better yet, the concert for SARS in Toronto in 2003, where the ratio of audience to performers was something like 100,000 to 1: a miniscule stage in a sea of people. That fits better with the point that Baars has repeatedly made, that there is a fundamental contradiction going on here. On the one hand, there are vast resources of the unconscious, working simultaneously—that is to say, in parallel—but those are forced to pass through a very narrow bottleneck to make it into the narrow, serial stream of consciousness. Here's an example: English is full of words with multiple meanings, most of which we know but only one of which we can conjure up at any moment. So, for instance, Baars points out that the word *set* can mean all kinds of different things, depending on the context: "jet set," "sunset," "ready, set, go," "a set of tennis," "a set of tools." Hearing only the word *set* may

---

* Experiments have demonstrated that we can remember literally thousands of visual images after seeing each only once for five seconds. Experts in visual memory have estimated that if we were shown a million pictures, we could recognize 986,000 of them if we were tested immediately.

incline you to one of those, or maybe none at all, but all are at the ready in the unconscious: add any of the context-setting words (which were not in your consciousness until you read them here) and the meaning immediately settles into place. That, of course, is the key to many jokes, where one meaning of a word is assumed until the punchline makes it clear that another was meant.

Once through the bottleneck and onto the stage, the stuff of consciousness is immediately fed back to the audience; consciousness allows information to be shared among disparate and presumably far-flung parts of the brain in a way that is assumed to be much more rapid and efficient than if those unconscious modules had to take care of it themselves.*

Not only is this theatre metaphor daring because it recalls the so-called Cartesian theatre that consciousness researchers love to ridicule, it also can be seen to rehabilitate the precise target of that ridicule, the homunculus who sits in the theatre and watches the drama of consciousness unfold before it. That homunculus is a problem because if it is taking it all in, you haven't solved the problem at all: you've just shrunk yourself down, and now you're faced with explaining how the homunculus is conscious, and so on and so on. But in fact, global workspace makes the homunculus respectable by increasing its numbers but decreasing its intelligence. The audience in this theatre can be envisioned as thousands of fairly stupid homunculi, each one capable of one or two simple bits of processing. None of them is conscious, and as long as their responsibilities are minimal, they lend themselves to scientific explanation.

One key to this setup is to ensure that all parts of the brain are easily accessible by consciousness. Given that there might be 100 billion neurons involved, this sounds like a daunting task, but it is also true, as I mentioned earlier, that any neuron can be connected to any other by no more than seven steps. That, of course,

---

* This is more than an assumption—it is the reason that global workspace architectures were created by theorists working in artificial intelligence in the first place: the best way to create appropriate problem-solving information from vast resources is to use a narrowly focused, short-lived, active method of selection.

is immediately reminiscent of the idea of six degrees of separation, that you can be connected to anyone else in the world through a chain of no more than six people. In 1998, mathematicians Steven Strogatz and Duncan Watts published an analysis of the kind of networks that might make such relationships possible, and showed that by combining the attributes of an "ordered" network (one in which every individual makes the same number of connections with others, all tightly clustered around each other) and a random network (connections all over the place), they could create a network that was mostly ordered but that had occasional long-range connections to other clusters.[4] This network mimicked the effects of the "six degrees of separation" model: it established that most relationships are short-distance, both geographically and socially, but that everyone has an occasional long-range acquaintance. While it has never been shown that this is what happens in the brain, such an arrangement would allow the connectivity between neurons that exists, and would also provide an unexpected link between the organization of the brain and that of human social life.

There is one thing to be cautious about here, and that is the risk of confusing your image of the global workspace with what it really must be like, if indeed consciousness is generated in this way. For one thing, both the theatre metaphor and the word "work*space*" imply a vast empty space in the brain, which of course is completely wrong. You can't fit hundreds of billions of cells into 1,400 cubic centimetres of tissue and have much empty space left over. The other misleading term is "module," which suggests—at least to me—a compact, well-defined unit, something you could plug into your computer. But in this case "module" really refers to a circumscribed function, not a place; something like the fusiform face area, which, while it might be confined to the temporal lobe, could easily include myriad neuronal extensions winding around each other and making connections to distant locations in the brain.

Also, the metaphor of a theatre is not intended to be taken literally in the sense that there is one circumscribed place, the stage, where consciousness takes place. The way it looks right now is that consciousness is not likely to be located in any one specific place, a site where it all comes together. There are parts of the brain thought to be crucial for it, but it doesn't necessarily reside in any one of them. So the theatre idea is a metaphor more of process than of structure. Whether the stage is the entire brain or one place at this moment and another at the next—or no place at all but a synchronicity—is not important.*

In addition, some commentaries on the theory have noted that the theatre metaphor may make sense from the point of view of process, but it is just too static to be a fair representation of consciousness. The fact that new ideas flash into consciousness every few tenths of a second contrasts dramatically with the action on a typical stage, and the separation of audience from stage and spotlight ignores the incredible fluidity of consciousness. In fact, the audience become actors, actors turn into audience, the spotlight moves crazily about and it's hard to tell the difference between the stage and the backstage. Sometimes there *is* no difference. Maybe most important to bear in mind is the fact we are creating the whole thing—actors, audience, stage, everything—from moment to moment. We are not observers in this theatre: we *are* the theatre. Metaphors are helpful, but they are definitely not reality.

But What's the Evidence?

Global workspace theory includes many elements, all of which can be matched more or less accurately to a theatre metaphor: a stage, a spotlight, a rapid succession of players on that stage, unseen

---

* However, there are many bold scientists who make arguments for certain places in the brain to be the seat of consciousness. For instance, physicist-mathematician John Taylor makes claims for the parietal lobes, on the basis that they house the body image and are the likely location for key parts of working memory. There is also evidence that visual consciousness involves not just visual areas of the brain, but parts of the frontal lobes as well, and most agree that the thalamus is another key element. At this point, theories are inclusive, not exclusive. John G. Taylor, "The Importance of the Parietal Lobes for Consciousness," *Consciousness and Cognition* 10 (2001): 379–417.

influences backstage, an unconscious audience, and most important, the dissemination of information from the stage to members of the audience, who otherwise would be oblivious to that information. Now, what evidence is there that this might represent an accurate picture of the mechanism of consciousness?

No one would argue with the claim that most of the brain is concerned with the unconscious. Some of that is entirely automatic processing, such as the transformation of patterns of light on the retinas of the eyes into recognizable objects (especially the independent and separate paths through which colour, motion and form travel). We can never experience any of that—how bewildering that would be!—probably because there are no circuits of neurons that can bring those processes into communication with the myriad other places in the brain necessary to raise them to consciousness. They are the equivalent of country roads with absolutely no connection to the freeway.* But there is also content in your brain that is unconscious *at this moment*, such as personal memories that you are not right now thinking about, which can be moved into consciousness when you want. There is also loads of evidence that this unconscious content influences what you think about and the decisions you make without ever having to enter consciousness (see Chapter 5 for examples).

There really isn't much debate over the idea that the cast that occupies the stage at any moment—that tiny percentage of what's in the brain that we are actually aware of—changes with incredible rapidity; you know from moment-to-moment experience that this is the case. You also know that whatever you're conscious of right now can cause other, up-until-now-unconscious knowledge to spring into awareness, populating the stage with a whole new set of actors. But is there anything beyond what it feels like to back up

---

* There has been speculation that those rare individuals called "autistic savants," some of whom can play a piece of music after one hearing or identify the day of the week of any date in the last half-century or conjure up prime numbers several digits long, are somehow accessing brain operations that what would be beyond reach for the rest of us. Of course, these individuals pay a price for those abilities: they are usually unable to cope with the everyday demands of life.

the argument that whatever is on the stage at this moment is being broadcast all over the brain to the larger, unconscious audience?

Two kinds of evidence suggest that this might be true. One is anatomical: the brain is wired so that activity in local areas can be communicated to distant locations and back again. There are sheaves of long-range neurons that run across the brain, left–right, from one hemisphere to the other, and front–back, from parietal to frontal lobes and back again. Prominent among these circuits are those that connect the thalamus, in the middle of the brain to the cerebral cortex, above, and back, and those connections play an important role in some theories of how consciousness works. It's possible to find neuron bundles connecting all the parts of the brain that would likely be required to maintain consciousness, including sensory areas, long-term and working memory sites, attention and motor areas. But the mere fact that there are connections in place doesn't prove that consciousness is a process that actually is using them to move signals around in the brain, sharing the short-lived contents of consciousness with many locales at the same time. You want to see those circuits in action, distributing information in a way that would be consistent with global workspace theory, and there is some evidence that this happens.

E. Roy John of the New York University School of Medicine has been using the EEG to probe the workings of the brain for decades, and in the late 1990s he and his colleagues designed a fascinating experiment in which they tracked the electrical activity of people's brains as they performed simple visual tasks that taxed their working memory.[5] A typical task was sixteen seconds long, consisting of a sequence of four seconds of looking at a simple cross on a screen, then four seconds of single letters, four seconds of the simple cross again, and finally four more seconds of letters, some of which had been present in the first set, some of which hadn't. In this example, subjects had to indicate whether each letter in the last set was present the first time letters had been shown or not.

The experiment let John predict what might appear: there should be some basic brain activity in response to the crosses, then some significant differences should appear as the subjects took in the first four seconds' worth of letters. The second appearance of the cross should be different from the first, because working memory was still active, holding on to the letter set (whereas the first time around it wasn't doing anything), then the final presentation of the letters should arouse new activity as that set was compared to the original set in working memory.

The results were intriguing. When the incredibly complex statistical processing was completed, allowing the researchers to tease what seemed to be the significant electrical signals (John called them "landscapes") out of the welter of background activity, three patterns emerged. The first was a slowly spreading wave of electrical activity beginning at the back of the brain and moving forward. This was exciting because it seemed to portray an initial registration of the image in the visual cortex at the back of the brain, followed by the forward movement that could reflect the further processing of that image by other centres in the brain. It is a particularly interesting sequence of events, because the frontal areas are known to be involved in choosing courses of action and planning, and seeing this sweep from back to front reminded the eminent brain scientist Robert Doty of the old neuroscientist's aphorism that the back of the brain is the past and the front is the future: the back receives the sensory input that reports on what's happened, then that information is forwarded to frontal areas where the response to those events is prepared.[6]

The second landscape revealed by John's experiment was a widening circle of activity, like ripples from a stone thrown in a pond, originating roughly in the centre of the brain. The third was activity that spread from the left hemisphere to the right. In a commentary on the results, Bernard Baars—perhaps hopefully—identified the second landscape as being centred on the thalamus, which of course is widely assumed to be crucial to distributing consciousness signals throughout the brain. The left–right activity suggested contributions from language areas moving to the other hemisphere. In a way, the precise patterns weren't as important as the fact that there *were* patterns— dynamic patterns, EEG signs of the broadcasting of information from an original source to distant locales in the brain.*

The theatre metaphor of the global workspace also includes a backstage composed of entities that somehow shape the context of consciousness. It sounds vague, but there is a startlingly clear example of it in the syndrome known as neglect. People who suffer from neglect have lost awareness of part of the space around them. It is almost always the space to the left that vanishes, as a result of damage to the right parietal lobe, the area of the brain at the top of the head. This part of the brain apparently maintains a map of the space around our bodies and, more important, makes that map available to consciousness. When the map is damaged, patients with the resulting neglect aren't aware that a portion of space has been obliterated; instead, they are unaware that there even should be such a space, and in most cases can't even remember that it once existed.

The most striking example of neglect, even though the syndrome has been researched thoroughly in the last twenty years, comes from the late 1970s. It was an experiment in which the subject matter was the cathedral square, the Piazza del Duomo, in Milan.[7] The square is dominated by the cathedral, built of white and pink Italian marble, containing thousands of statues and hundreds of spires. The square in front of the cathedral is about the size of a football field, empty except for a statue of King Victor Emmanuel urging his troops on in battle. There are buildings all around the square, including the glassed-in nineteenth-century shopping mall called the Galleria, the Duomo hotel, the department store Rinascente and the men's clothing shop Galtrucco. Why was this a good setting for an experiment? Because everyone

---

* Funnily enough, John himself, while acknowledging that identifying these electrical trends in the brain could lend support to the idea of the global workspace, has since published his own, different, theory of consciousness.

who lives in Milan knows the cathedral square intimately. Neuropsychologist Edoardo Bisiach asked two of his Milanese patients who had suffered strokes to the right side of their brains to envision the cathedral square and list as many of the buildings as possible. The trick was that they were to imagine two different points of view: first, standing at the end of the square looking towards the cathedral, then reversing that by imagining being on the steps of the cathedral looking back.

Both patients were able to name several of the buildings on the right as you look at the cathedral but none on the left. However, when asked to imagine standing on the cathedral steps, they immediately came up with a completely different list that included the buildings now on the right (buildings they had failed to name before) and none of the buildings now on their left, buildings they had just finished naming. The inability of these two patients of Bisiach's even to *imagine* the left side of the space around them demonstrated that neglect was a matter of an inability not simply to see anything off to the left, but to be *aware* of it. With some persuasion one of the patients was able to name a single building on the left, but for the most part the patients were completely consistent in being unaware of the left side of the cathedral square, no matter what their imagined perspective.

To come back to the global workspace idea, think of the right parietal lobe as one of those context-setting agents backstage, which, while not actually contributing any lines of script, nonetheless make what is happening on the stage intelligible by, in this case, providing an awareness of the space around us. It's the difference between the intact right hemisphere knowing something is wrong, say, with the left arm, and reporting that, and the right hemisphere being damaged and unable to report anything. This example does double duty in that the critical deficit in neglect involves the inability to pay attention to the left side of space, and, of course, without the spotlight of attention focused on it, no area of activity in the brain can enter consciousness.

So far, then, there is some evidence that consciousness might depend on the brain's being organized as some sort of global

workspace. As I said, this is only one example of a model of consciousness that has been put on the table, and while it has had
reasonable success in explaining how consciousness might work,
not every expert buys it. There is no single criticism that stands
out, but one in particular gives a flavour of the kind of objections
that can be raised. Dr. Susan Blackmore thinks that the global
workspace idea is a nonstarter. Blackmore, a psychologist and
writer, is completely unconvinced by the theory because it
implies a stream of consciousness, an unfolding sequential story
on the stage—something she thinks simply doesn't exist.[8]

She argues this way: global workspace theory implies that
things are either in or out of consciousness, but there are experiences that argue against that. For instance, imagine, she says, that
you're reading at eight o'clock at night and you suddenly become
aware of a sound, like the chiming of a clock—something that, as
far as you know, you were unaware of until this moment. If that
were true, how is it possible that you can now be aware that the
clock has already struck four times and will strike four more. The
sequence of chimes must have entered your mind somehow, and
yet as far as you can tell, that sequence doesn't seem to have
been part of your stream of consciousness.

I'm not sure I have ever experienced something exactly like
this (although I have talked to people who have), a moment
where a sudden awareness of something continuous can be
backdated to the point where it started, but this particular example leads directly to a theory that Blackmore espouses, which
she contrasts with the stream of consciousness by calling it an
"illusory backwards stream." She contends that there really is no
single definable thing in consciousness at any time; it just seems
that way when you probe it. So, for instance, someone asks you
what you were thinking about, or something catches your attention and motivates you to think about what was going on in your
mind. In that circumstance it's easy to come up with a notion of
what was in your consciousness, but what was going on in the
absence of such a probe? Blackmore argues that your brain is

juggling many different bits of information, feelings, perceptions, memories, in no particular order. It's the brain's version of
quantum theory, where events are indeterminate until probed
by the instruments of the physicist, and only then do the multiple possibilities resolve themselves into one. Blackmore acknowledges a resemblance here to Daniel Dennett's famous multiple
drafts theory, in which the arrow of time and the stream of
consciousness give way to a confused simultaneity, a bustling
marketplace of bits of information, none of them in boxes
labelled "conscious" and "unconscious." At least, none of them
are in those boxes until we claim that they are: that's the difference between an actual well-defined sequence of mental events
and an impression of one.

It's clear that Blackmore views the division between conscious
and unconscious with suspicion. In her mind there is no division
(and I mean that in both senses in which it can be taken), no
stream that can be identified as conscious as opposed to the rest,
but only the retrospective appearance of one. Blackmore tells us
that she prompts her students to ask themselves, many times a day,
"Am I conscious now?" She says that with repetition, such self-
examination at first creates confusion, then perhaps reveals many
threads—a plurality of sounds such as the chiming clock would be
an example—that overlap, intertwine and perhaps represent several mini-streams of consciousness, not just one. Their existence is
not independent, but is created by the act of reflection.

Returning for the last time to the now well-worn example of
highway driving while unconscious, I can see how Blackmore's
idea might apply: it isn't that you actually drove those last ten
kilometres unconsciously, but rather, now that you're looking
back on them, it *seems* you did, because you can't remember paying attention. But couldn't it be possible that you actually were, at
least every few moments, then returning to what was primarily
occupying your mind—the radio or your passenger? If those brief
conscious moments weren't memorable, then it would seem as if
you hadn't been paying attention at all.

But how can Blackmore's contention be proved? If, by definition, consciousness is personal and private (even brain imaging that reveals activity doesn't show whether that activity is conscious or not), and if every time I perform the act of inspecting it I discover a stream, how can it ever be established that the stream is an illusion? Blackmore admits this is a little like flinging the refrigerator door open as quickly as you can or as sneakily as possible to see if the light is always on, but she maintains that such reflection, especially when it suggests a multiplicity of thoughts prior to checking them out, backs up her contention that there never was a single stream.

I'll admit I haven't practised this a lot, but from my personal point of view, I feel as if have no trouble monitoring an ongoing sequence of thoughts, feelings, memories, even weird associations. I have to admit though that there is one moment during which I have trouble identifying what I am conscious of, and that's when I'm speaking. If you figure out what you want to say then say it, what's going on in your mind as you actually utter the words? In my case it appears not to be a lot—or maybe even nothing. That may sound like a direct contradiction of the old saying popular among neuroscientists, "How do I know what I'm thinking until I say it?"—implying that thought is actually bound up so completely with speech that you can't claim that thought comes first—but I don't think it is. When you speak, you do have the thought you want to express, but usually not the precise set of words to express it. Those seem to come out more or less spontaneously. Even so, when they're being uttered, what's happening in your mind? Try it.

Bernard Baars is not deterred by Blackmore's position, and commented to me by e-mail that "if Sue Blackmore is right, evolution has done us a dirty deed. We could not predict the future consciously. But we plainly do so every day, or we would end up wandering aimlessly through the world, just tasty snacks for any passing tiger. So the hypothesis strikes me as wrong based on our everyday experience. If we can't predict the immediate future (using both conscious and unconscious knowledge) what good are we?"

It's impossible to say at this point how this debate will be resolved: as I said, Susan Blackmore's doubts about global workspace theory represent only one approach to the matter. I haven't actually forced myself to count the number of theories of how consciousness works, but there must be dozens. But I would be willing to bet that somewhere between the theatre metaphor and the disordered chaos of models like Blackmore's lies the truth, at least as far as it will be possible to determine it.

It wouldn't make sense to leave this topic without pointing out how much more needs to be explained. Assume for the moment that global workspace is essentially correct; what does it leave out? Start with the idea that consciousness makes information available to the unconscious modules in order to bring all possible information to bear on the situation. How exactly does that broadcasting occur? Most believe that there has to be some sort of synchronizing of the firing of neurons in parts of the brain that are sharing information, but does that mean that the neurons are firing at exactly the same time, or that they've coordinated their rate of firing, so that the slower ones have speeded up and faster ones slowed down? If it's rate of firing that creates consciousness, does it come into existence abruptly, as the intensity of neural activity crosses a critical threshold, or gradually, as that activity ramps up? All these versions are on the table.

Try to pull something into your consciousness right now, such as the first family car you can remember. What happened in your brain as that internal image came into view? Was it the intensity or rapidity of a certain set of nerves firing, so that what was out of mind became the focus of your thoughts? Remember that brain imaging has revealed that certain areas are intensely active in concert with awareness of what those areas are doing, such as the fusiform face area and the mental image of a face. But those areas can be active in the absence of consciousness too. What makes the difference? And what role do the neurotransmitters play in all this? They are essential for a nerve impulse to pass from one neuron to another, and considering that single neurons can sometimes make thousands of connections with others, you can see that the chemistry of consciousness is hugely important. However, with the singular exception of dreaming, not much is known about it. It certainly isn't clear what the transmitters had to do with remembering that family car, but it is perfectly clear that in situations where neurotransmitters or their receptors are deranged, such as schizophrenia, consciousness is altered.

While these are all important problems to be solved, the ultimate one, the hard problem, remains not just unresolved, but unapproached, not just by global workspace theory, but by all the others.* Just think back to the family car example (which by now would likely have lapsed back into some place in your unconscious) and ask yourself what exactly you remember about it, besides the obvious things like colour and model? Is your mental image of the car as seen from standing outside it, or the view from the seat where you usually sat, or the "clunk" of the car doors, the new car smell or the arguments your mother and father had in the front seat? Now ask yourself, how exactly does some sort of heightened activity in multiple places in the brain actually *create* those images and the feelings that flood billions of synapses? The attentional spotlight may be wandering around, but that is all physics and chemistry, not mentality. Thoughts have no substance, or at least no substance that we have yet discovered. The hard problem, as philosopher David Chalmers called it, is producing those thoughts from the chemistry and physics that we're familiar with.**

---

* The proponents of quantum theories of consciousness would argue that theirs are the only theories that make the hard problem accessible, by turning regular everyday chemistry and physics into quantum phenomena. Doubters dismiss these claims by saying that the only thing quantum physics and consciousness have in common is that they're both mysterious.

**Chalmers has resorted to arguing that consciousness is one of the fundamental properties of the universe, like gravity and the charge on the electron. If so, it's no surprise we haven't been able to come to grips with it yet.

Of all the issues in consciousness that divide scientists from philosophers and that create splinter groups within each discipline, the hard problem is prime. It discourages some from even attempting to make progress; it drives some out to the very frontiers of science—and beyond—to try to narrow the gap between the phenomena that need to be explained and the data available; and it leaves others having to admit that while they're not really equipped to come up with an explanation right now, they see no point in abandoning the problem because of that. But with new technology, knowledge and years of research, the answers that seem so distant and so foreign right now may someday emerge.

Epilogue

I BEGAN this book with the fear of writing a book about a subject nobody could define; I end it with another fear: that there will have been larger-than-life questions raised in your mind by what you've read that unfortunately are never answered. I'd (seriously, not sarcastically) recommend something like the *Oxford Companion to the Mind* for the answers to those big questions. There's a time and place for everything, and I set out to paint a picture—as well as I could—of the consciousness that most us share. But it wouldn't be right to leave without hinting at how everyday consciousness can, in the right circumstances, morph into something spectacularly different.

"Cosmic consciousness" definitely qualifies as one of those sensational variations. It probably comes in several different kinds and can be achieved in different ways, but it is in general a brief time during which you have feelings of calm and elation, a feeling of oneness with god, with the afterlife and the universe, and other features impossible to describe with words—they are, as the philosophers like to call them, "ineffable."

Two pieces of personal testimony provide a sense of the cosmic experience. Personal accounts are important in consciousness studies, as we've seen, but are much more important in cases like these, where the conscious experience is so different from that of most people. The first was written by the nineteenth-century Canadian physician Richard Maurice Bucke, founder of the University of Western Ontario's medical school, and a most unusual guy. Before he was even twenty-one he had already ridden with a wagon train out of Fort Leavenworth, searched for silver in California and lost a foot to frostbite. He spent much of his life practising medicine in western Ontario, but travelled often to England—where he had had some schooling—and it was on a trip there that he had a life-changing experience.

He describes the event in the third person, but the man he writes about is himself. He was riding home in a hansom cab in London in the late 1800s after an evening of reading poetry with his friends, when,

> All at once, without warning of any kind, he found himself wrapped around as it were by a flame colored cloud. For an instant he thought of fire, some sudden conflagration in the great city, the next he knew that the light was within himself. Directly afterwards came upon him a sense of exultation, of immense joyousness, accompanied or immediately followed by an intellectual illumination quite impossible to describe.
>
> He saw and knew that the Cosmos is not dead matter but a living Presence, that the soul of man is immortal, that the universe is so built and ordered that without any peradventure all things work together for the good of each and all."[1]

Bucke's transcendental experience lasted only a few moments, but never really left him; he remembered it always, and never doubted that it had actually happened to him.

A more recent perspective was provided, in the same article in the *Journal of Consciousness Studies*, by Allan L. Smith. A scientist, Smith's life was forever changed by an experience that in many ways seems similar to Bucke's. Smith then went on to take several acid trips, in the hopes of duplicating the spontaneous experience he had had. While he wasn't successful (and was clearly disappointed by that), he was at least in a position to differentiate the two kinds of experience.

Smith's cosmic moment started in a most peculiar way: he was sitting alone in an easy chair, watching the sunset, when he felt an unusual tingling in his perineum, the region between the genitals and anus (or, as my wife suggested putting it, "the neutral place between the front and the back"). Just exactly how that presages something cosmic is for you to figure out. The tingling soon gave way to changes in the surrounding light, and then something more profound: "I merged with the light and everything, including myself, became one unified whole. There was no separation between myself and the rest of the universe. In fact, to say that there was a universe, a self, or any 'thing' would be misleading. . . . In fact, there were no discrete events to 'happen' — just a timeless, unitary state of being."

In Smith's case, the ecstasy lasted a good twenty minutes (as far as he could tell, given that time had apparently stood still). He, like Bucke, sought to repeat it and never was able to do so, but he also tried those acid trips in an attempt to duplicate it. His LSD experiences shed some interesting light on the differences between the two. Some features were similar: time slowed, light changed and, as he put it there was sometimes "an experience of a wordless significance of certain objects."

However, in every respect the acid version fell short of the cosmic: time slowed, but it didn't stop; his mood was brittle and changeable, not one of elation; his self may have been temporarily displaced, but it was never lost entirely. Smith is convinced that the two experiences, while superficially similar, are actually two different states of consciousness. If you add those to dreaming and waking consciousness, the catalogue of human versions begins to swell.

What could be happening here? One of the interesting things about these personal feelings is that at least one of them is also experienced in certain kinds of meditation.

Andrew Newberg of the University of Pennsylvania, author of the book *Why God Won't Go Away*, has used brain imaging to look at

the activity in the brains of experienced Tibetan Buddhist meditators and has found that when they meditate, their frontal lobes are
more intensely active but the parietal lobes virtually shut down. It's
tempting to correlate these results with the experience of meditation. The frontal lobes enable you to concentrate intensely, which is
exactly what you might do when meditating by repeating a simple
mantra. One of the important roles of the parietal lobes is to establish the self, its position and place in space, and the shape of space
around it. You could argue, as Newberg has, that the brain changes
you'd expect to see in someone who feels as if the boundary
between the self and the neighbouring space has disintegrated are
exactly what *is* seen: diminished activity in the parietal lobes, the
very places where the separation is created in the first place.*

It sounds perfect, right? You are able to lose yourself in the
universe only because your brain takes down the boundaries that
usually prevent that from happening. You have to be careful in
jumping to such conclusions, though: damage to the same lobes
can create neglect, that state of mind in which the left side of
space is left outside consciousness. In the cosmic experiences,
there is apparently no loss of awareness of space, just a change in
its nature. Anyway, as I'm sure you know already, we're just at the
beginning of being able to explain exactly what was going on
there in the parietal lobes with respect to consciousness.

There is one curious link here to other aspects of consciousness.
I received an e-mail from a practitioner of *surat shabd* yoga, who
described a similar withdrawal from the physical body: "This is a
very real sensation. It begins at the extremities, proceeds up the
limbs, on up the body proper, and ends with the feeling of being
more or less a point of consciousness, centered between the two
eyes at the upper base of the nose. The eyes are kept closed, with
the gaze centered in the center of the field of view." What's striking

---

* Curiously, a recently published study of Tibetan monks showed that when
viewing an ambiguous figure like the Necker Cube, the monks were able to prevent the image from switching from one version to the other, sometimes for
minutes at a time.

about this account is not only the dissolution of the body, but the fact that what remains is a point of consciousness located at exactly the very same place—between the eyes and behind the nose—that consciousness seems to occupy even in ordinary waking.

While Allan L. Smith found that life on LSD paled in comparison to his cosmic experience, there have been other reports of drug-induced changes to consciousness that suggest that more, and different and weirder, states are possible.

You might know Dr. Andrew Weil best from his smiling face on billboards advertising vitamins or his cookbook, but Weil has an intriguing history of investigations into consciousness, and he contributed a fascinating account of one of them to a collection of consciousness essays.[2] Weil is both a botanist and a doctor, and in his early career he spent time investigating the mind-altering properties not just of relatively standard drugs like marijuana and psilocybin, but of exotic drugs used by tribes inhabiting the Amazon basin. The most striking thing about some of these drugs are the reports that, unbelievable as it may seem, people taking them in a group somehow share the same conscious experiences as the people around them.

Weil himself describes an experience he had smoking the prepared venom of a desert toad, a drug that in his estimation would be a little too frightening for most people inexperienced with drug-taking. For many, the initial impact of the venom is a loss of awareness for five or seven minutes—the drug-taker simply isn't there any more. Then, gradually, normal awareness returns, and some users have described the return as the fascinating part (and I'm sure reassuring) part. Weil did it once: "I inhaled the vapors with a friend who had used it before, and the initial phase of the intoxication was—well, I just do not remember it. But in the period of reconstituting and becoming aware of myself and external reality, there was a distinct experience that lasted for a minute or two, in which we simply seemed to be present in each other's consciousness. When we were able to talk again, my friend told me he had had the same experience."

This may be a remarkable, even freaky experience, but it's just a taste of what others have reported. Weil quotes the experience of a man named Manuel Curdova sharing a drug called ayahuasca with a group of twelve other men, all of whom, once intoxicated, experienced visions of a series of strange jungle animals: "Some of these I had not seen before. There was a tawny puma, several varieties of the smaller spotted ocelot, then a giant rosetta-spotted jaguar. A murmur from the assembly indicated recognition. This tremendous animal shuffled along, head hanging down, mouth open and tongue lolling out. Hideous large teeth filled the open mouth. An instant change of demeanour to vicious alertness caused a tremor through the circle of phantom viewers."

What could be better than mysteries piled on top of mysteries? Ordinary consciousness has innumerable features that cannot yet be explained, features that the people who experience them cannot even agree on. And now shared consciousness? Weil ends his essay with this: "I do not think that we can eliminate mystery from our experience of reality, nor is it the business of science to try to do that. Science can approach mystery. But it seems that the brighter we illuminate reality with the light of science, the more we become aware of the extent of the surrounding darkness."

Understanding the routine of daily waking consciousness is mysterious enough; as I've said more than once, there are many who are convinced it will never be explained. But if scientists don't at least try to pull back the curtain a little, to get a glimpse into what must be the wondrous mechanics of it, how can we even appreciate, let along understand, stories like the ones I've quoted here?

I think all consciousness scientists would acknowledge that there is no respectable theory anywhere that could even begin to explain how a group sitting around a campfire could "see" the same threatening animal. But that doesn't mean we won't one day have some insights into that. But first, we need to understand the simpler things, like deciding right now that what you'd like more than anything is a beer. Go for it! But don't assume your conscious mind had anything to do with it.

Acknowledgments

W RITING this book has made one thing clear to me: being an author's researcher must be one of the most frustrating jobs on earth. In this case, Tamara Adamek was the "lucky" person. Her contributions to the finished product were likely not what she had envisioned: some of the carefully selected (but exhaustive) binders of papers that she provided on the various topics I had suggested still make a tidy pile in my office, neglected but not unappreciated. It wasn't her fault: I just kept changing my mind about what was to go in the book and what wasn't. What she did contribute, however, was unfailing enthusiasm, interest, incisive critical comments, a sense of humour and a belief in the book. If that's what research turns out to be, fine. I couldn't have made progress without it.

Kevin Hanson of HarperCollins shared that undying faith. He practically invented the book and has never doubted it. Jim Gifford actually read the entire manuscript more than once (courageous man) in the course of editing it. Copy editor Stephanie Fysh turned the prose into English with a unique combination of attention to detail and humour.

Many scientists responded to requests for information and for pre-publication versions of scientific papers, including Bernard Baars and his colleagues at the Neurosciences Institute in San

Diego, Christof Koch, James Rutka, Ron Rensink and Brock Fenton (who described life as a bat). I thank them. But two psychologists — Alan Kingstone and Allison Sekuler — have influenced and encouraged me more than any others, during both the preparation of this book and beforehand, and I thank them especially.

Endnotes

Chapter 1

1 J.M. Evans, "Patients' Experiences of Awareness during General Aneasthesia," in *Consciousness, Awareness and Pain in General Anaesthesia*, ed. M. Rosen and J.N. Lunn, (London: Butterworth's 1987): 184–92.

2 F. Jackson, "Epiphenomenal Qualia," *Philosophical Quarterly* 32 (1982): 127–36.

3 D. Chalmers, "Facing Up to the Problem of Consciousness," *Journal of Consciousness Studies* 2 (1995): 200–19.

4 Christof Koch, *The Quest for Consciousness: A Neurobiological Approach* (Englewood, CA: Roberts, 2004): 238.

5 Richard Gregory, "Flagging the Present Moment with Qualia," in *Toward a Science of Consciousness III*, ed. Stuart R. Hameroff, (Cambridge, MA: MIT Press, 1998): 259–69.

6 V.S. Ramachandran and W. Hirstein, "The Three Laws of Qualia," *Journal of Consciousness Studies* 4 (1997): 429–58.

7 John McCarthy, *Ascribing Mental Qualities to Machines* (1979), http://www-formal.stanford.edu/jmc/.

8 Norman Maier, "Reasoning in Humans: The Solution of a Problem and Its Appearance in Consciousness," *Journal of Comparative Psychology* 12 (1931): 181–94.

9 Timothy Wilson and Richard Nisbett, "The Accuracy of Verbal Reports About the Effects of Stimuli on Evaluations and Behavior," *Social Psychology* 41 (1978): 118–31.

Chapter 2

1 T. Koenig, K. Kochi and D. Lehmann, "Event-Related Electric Microstates of the Brain Differ Between Words with Visual and Abstract Meaning," *Electroencephalography and Clinical Neurophysiology* 106 (1998): 535–46.

2 William James, *The Principles of Psychology* (New York: Henry Holt, 1890).

3 Transcripts collected by Kenneth S. Pope and published in *The Stream of Consciousness: Scientific Investigations into the Flow of Human Experience*, ed. Kenneth S. Pope and Jerome L. Singer (New York: Plenum, 1978).

4 James, "The Stream of Thought," in Chapter 9 *The Principles of Psychology*.

5 Alain Morin, "Inner Speech and Conscious Experience," *Science and Consciousness Review* 4 (April 2003), http://www.sci-con.org.

6 Blackmore is notable because she experienced a vivid OBE herself, in which she was suspended above her body and was able to watch her mouth open and close as she spoke. In Susan J. Blackmore, *Beyond the Body: An Investigation of Out-of-the-Body Experiences*. (London: Heinemann, 1982) p. 165.

7 Julian Jaynes, "Consciousness and the Voices of the Mind," *Canadian Psychology* 27 (1986): 128–39.

8 Charles Tart, "Psychophysiological Study of Out-of-the-Body Experiences in a Selected Subject," *Journal of the American Society for Psychical Research* (1968): 3–27.

9 Charles Tart, "Six Studies of Out-of-the-Body Experiences," *Journal of Near-Death Studies* 17 (1998): 73–99.

Chapter 3

1 *Cranial Pursuits*, with Ira Basen, Christopher Grosskurth and Ben Schaub, CBC Radio.

2 John McCrone, "Exploding the 10 Percent Myth," *Science and Consciousness Review* 1 (2004).

3 Dale Carnegie, *How to Stop Worrying and Start Living* (New York: Simon and Schuster, 1944), 123.

4 J. Zihl, D. Von Cramon and N. Mai, "Selective Disturbance of Movement Vision after Bilateral Brain Damage," *Brain* 106 (1983): 313–40.

5 Sandra Witelson, Debra Kigar and Thomas Harvey, "The Exceptional Brain of Albert Einstein," *The Lancet* 353 (1999): 2149–53.

6 Terence Hines, "Further on Einstein's Brain," *Experimental Neurology* 150 (1998): 343–44.

Chapter 4

1 Daniel Dennett, "Julian Jaynes's Software Archeology," *Canadian Psychology* 27 (1986): 149–54.

2 John Eccles, "Do Mental Events Cause Neural Events Analogously to the Probability Fields of Quantum Mechanics?" *Proceedings of the Royal Society of London, Series B, Biological Sciences* 227, no. 1249 (1986): 411–28.

3 Pim van Lommel, Ruud van Wees, Vincent Meyers and Ingrid Elfferich, "Near-Death Experience in Survivors of Cardiac Arrest: A Prospective Study in the Netherlands," *The Lancet* 358 (2001): 2039–45.

4 George Miller, "The Magic Number Seven, Plus or Minus Two: Some Limits on Our Capacity for Processing Information," *The Psychological Review* 63, no. 2 (1956): 81–97 .

5 B. Barrs, "A Biocognitive Approach to the Conscious Core of Immediate Memory," *Behavioral and Brain Sciences* 24 (2000): 115–16.

Chapter 5

1 William James, *The Principles of Psychology*, Chapter 26.

2 Julian Jaynes, *The Origin of Consciousness in the Breakdown of the Bicameral Mind* (Boston: Houghton Mifflin, 1976).

3 M. Goodale and D. Milner, *Sight Unseen: An Exploration of Conscious and Unconscious Vision* (Oxford: Oxford University Press, 2004).

4 P. Merikle, D. Smilek and J.D. Eastwood, "Perception without Awareness: Perspectives from Cognitive Psychology," *Cognition* 79 (2001): 115–34.

5 W.R. Kunst-Wilson and R.B. Zajonc, "Affective Recognition of Stimuli That Cannot Be Recognized," *Science* 207 (1980): 557–58. Robert Zajonc has been influential enough through his career to have merited the publication of his "selected works."

6 A. Marcel, "Conscious and Unconscious Perception: Experiments on Visual Masking and Word Recognition," *Cognitive Psychology* 15 (1983): 197–237.

7 Abraham Pais, *"Subtle Is the Lord . . . ": The Science and the Life of Albert Einstein* (Oxford: Oxford University Press, 1982): 179.

8 Janet E. Davidson, "The Suddenness of Insight," in T*he Nature of Insight*, ed. Robert J. Sternberg and Janet E. Davidson, (Cambridge, MA: MIT Press, 1996): 33–62.

9 K. Duncker, "On Problem Solving," *Psychological Monographs* 58, no. 5 (1945).

10 Mark Jung-Beeman, Edward M. Bowden, Jason Haberman, Jennifer L. Frymiare, Stella Arambel-Liu, Richard Greenblatt, Paul J. Reber, John Kounios, "Neural Activity When People Solve Verbal Problems with Insight," *PLOS Biology* 2 (4) (2004), http://biology.plosjournals.org.

11 John A. Bargh, "The Automaticity of Everyday Life," in *Advances in Social Cognition*, ed. R.S. Wyer Jr., vol. 10, (Mahwah, NJ: Erlbaum, 1997): 1–61.

Chapter 6

1 "Recovery from Blindness," *The Oxford Companion to the Mind*, ed. Richard Greg, (Oxford: Oxford University Press, 1991): 94.

2 R.L.G. Gregory and J.G. Wallace, *Recovery from Early Blindness: A Case Study*, Experimental Psychology Society Monograph 2, (1963; 2001). Oliver Sacks also described a similarly unhappy case in *An Anthropologist from Mars: Seven Paradoxical Tales* (New York: Vintage, 1995).

3 I. Fine, A.R. Wade, A.A. Brewer, M.G. May, D.F. Goodman, G.M. Boynton, B.A. Wandell and D.J. MacLeod, "Long-term Deprivation Affects Visual Perception and Cortex," *Nature Neuroscience* 6 (2003): 915–16.

4 D. J. Simons and C.F. Chabris, "Gorillas in Our Midst: Sustained Inattentional Blindness for Dynamic Events," *Perception* 28 (1999): 1059–74.

5 Kevin O'Regan, "Solving the 'Real' Mysteries of Visual Perception: The World as an Outside Memory," *Canadian Journal of Psychology* 46 (1992): 461–88.

6 W. James, *Principles of Psychology*.

7 R. L. Gregory, *The Intelligent Eye* (London: Weidenfeld and Nicolson, 1970), 37.

8 Jerome Lettvin, "On Seeing Sidelong," *The Sciences* 16 (1976): 10–20.

Chapter 7

1 Daniel Dennett, *Consciousness Explained* (Boston: Little, Brown, 1992).

Chapter 8

1 B. Libet, E.W. Wright Jr., B. Feinstein, and D.K. Pearl, "Subjective Referral of the Timing for a Conscious Sensory Experience: A Functional Role for the Somatosensory Specific Projection System in Man," *Brain* 102 (1979): 193–224.

2 D.C. Dennett and M. Kinsbourne, "Time and the Observer: The Where and When of Consciousness in the Brain," *Behavioral and Brain Sciences* 15 (1992): 183–247.

3 B. Libet, "Unconscious Cerebral Initiative and the Role of Conscious Will in Voluntary Action," *Behavioral and Brain Sciences* 8 (1985): 529–66.

4 Azim F. Shariff and Jordan Peterson, "Anticipatory Consciousness, Libet's Veto, and a Close-Enough Theory of Free Will," in *Consciousness & Emotion*, ed. Ralph D. Ellis and Natika Newton, (Amsterdam: John Benjamins, 2005): 197–215.

5 Benjamin Libet, in *Mind Time: The Temporal Factor in Consciousness* (Cambridge, MA: Harvard University Press, 2004).

Chapter 9

1 P. de Laplace, "Essai philosophique sur les probabilités," trans. F.W. Truscott and F.L. Emory as *A Philosophical Essay on Probabilities* (1820; New York: Dover, 1951).

2 W. Wegner and T. Wheatley, "Apparent Mental Causation: Sources of the Experience of Will," *American Psychologist* 54 (1999): 480–92.

3 Fritz Heider and Mary-Ann Simmel, "An Experimental Study of Apparent Behavior," *American Journal of Psychology* 53 (1944): 243–59.

4 Marvin Minsky, *The Society of Mind* (New York: Simon and Schuster, 1985).

5 R.F. Baumeister, E. Bratslavsky, M. Muraven and D.M. Tice, "Ego Depletion: Is the Active Self a Limited Resource?" *Journal of Personality and Social Psychology* 74 (1998): 1252–65.

Chapter 10

1 Thomas Nagel, "What Is It Like to Be a Bat?" *The Philosophical Review* 83 (1974): 435–50.

2 Carolyn Ristau, *Aspects of the Cognitive Ethology of an Injury-feigning bird, the Piping Plovers*, in *Cognitive Ethology: The Minds of Other Animals*. ed. C.A. Ristau (Hillsdale, New Jersey: Lawrence Erlbaum, 1991): 91–126.

3 Robert H. Wozniak, "Conwy Lloyd Morgan, Mental Evolution and the *Introduction to Comparative Psychology*," in C.L. Morgan, *Introduction to Comparative Psychology*, vii–xix (London: Routledge, 1993).

4 Irene Pepperberg and Spencer Lynn, "Possible Levels of Animal Consciousness with Reference to Grey Parrots (*Pstittacus erithacus*)," *American Zoologist* 40 (2000): 893–901.

5 A. Seth, B. Baars and D. Edelman, "Criteria for Consciousness in Humans and Other Mammals," *Consciousness and Cognition* 14 (2005): 119–39.

6 N.J. Emery and N.S. Clayton, "Effects of Experience and Social Context on Prospective Caching Strategies by Scrub Jays," *Nature* 414 (2001): 443–46.

7 Alain Morin, guest editorial, *Science and Consciousness Review*, no. 1 (November 2003).

8 R. Clark and L. Squire, "Classical Conditioning and Brain Systems: The Role of Awareness," *Science* 280 (1998): 77–81.

9 Daisie Radner, *Animal Consciousness* (Amherst, New York: Prometheus Books, 1989): 179–78.

10 Sue Savage-Rumbaugh, William Mintz Fields and Jared Taglialatela, "Ape Consciousness–Human Consciousness: A Perspective Informed by Language and Culture," *American Zoologist* 40 (2000): 910–21.

Chapter 11

1 Thomas H. Huxley, "On the Hypothesis That Animals Are Automata, and Its History," *The Fortnightly Review*, n.s. 16 (1874): 555–80.

2 Francis Galton, "Psychometric Facts," *Nineteenth Century* (March 1879): 425–33.

3 Julian Jaynes, *The Origin of Consciousness in the Breakdown of the Bicameral Mind* (Boston: Houghton Mifflin, 1976).

4 F. C. Crick and C. Koch, "Are We Aware of Neural Activity in Primary Visual Cortex?" *Nature* 375 (1995): 121–23.

5 Belinda R. Lennox, S. Bert, G. Park, Peter B. Jones and Peter G. Morris. "Spatial and Temporal Mapping of Neural Activity Associated with Auditory Hallucinations," *The Lancet* 353 (1999): 644.

6 S.S. Shergill, M.J. Brammer, E. Amaro, S.C. Williams, R.M. Murray and P.K. McGuire, "Temporal Course of Auditory Hallucinations," *The British Journal of Psychiatry* 185 (2004): 516–17.

7 Mark Leary and Nicole Buttermore, "The Evolution of the Human Self: Tracing the Natural History of Self-Awareness," *Journal for the Theory of Social Behavior* 33 (2003): 365–404.

8 Nicholas Humphrey, "Cave Art, Autism, and the Evolution of the Human Mind," *Journal of Consciousness Studies* 6, no. 6–7 (1999): 116–43.

9 Steven Mithen, "Handaxes and Ice Age Carvings: Hard Evidence for the Evolution of Consciousness" in *Toward a Science of Consciousness III: The Third Tucson Discussions and Debates*, ed. Stuart R. Hameroff (Cambridge, MA: MIT Press, 1999).

10 Bjorn Merker, "The Liabilities of Mobility: A Selection Pressure for the Transition to Consciousness in Animal Evolution." *Consciousness and Cognition*, vol. 14, (2005): 89–114.

11 Matt Rossano, "Expertise and the Evolution of Consciousness," *Cognition* 89 (2003): 207–36.

Chapter 12

1 Katherine Nelson, "Explaining the Emergence of Autobiographical Memory in Early Childhood," in *Theories of Memory*, ed. A.F. Collins, S.E. Gathercole, M.A. Conway and P.E. Morris, (Hillsdale, NJ: Erlbaum, 1993): 355–85.

2 G. Simcock and H. Hayne, "Breaking the Barrier? Children Fail to Translate Their Preverbal Memories into Language," *Psychological Science* 13 (2002): 225–31.

3 K. Nelson and R. Fivush, "The Emergence of Autobiographical Memory: A Social Cultural Developmental Theory," *Psychological Review* 111 (2004): 486–511.
4 Nicholas Humphrey, *The Mind Made Flesh: Essays from the Frontiers of Psychology and Evolution* (Oxford: Oxford University Press, 2002).

Chapter 13

1 Arthur Ladbroke Wigan, *A View of Insanity: The Duality of the Mind* (1844; Malibu, CA: Joseph Simon, 1985).
2 Victor Mark, "Conflicting Communicative Behavior in Split-Brain Patient: Support for Dual Consciousness," in *Toward a Science of Consciousness*, ed. S.R. Hameroff, A.W. Kaszniak and A.C. Scott, (Cambridge, MA: MIT Press, 1996): 189–96.
3 M. Gazzaniga, "Cerebral Specialization and Interhemispheric Communication," *Brain* 123 (2000): 1293–1326.
4 Alain Morin, "The Split Brain Debate Revisited: On the Importance of Language and Self-Recognition for Right-Hemispheric Consciousness," *Journal of Mind and Behavior* 22 (2000): 207–18.
5 C.S. Moss, *Recovery with Aphasia: The Aftermath of My Stroke* (Chicago: University of Illinois Press, 1972).
6 V.S. Ramachandran, L. Levi, L. Stone, D. Rogers-Ramachandran, R. McKinney, M. Stalcup, G. Arcilla, R. Zweifler, A. Schatz and A. Flippin, "Illusions of Body Image: What They Reveal about Human Nature," in *The Mind-Brain Continuum: Sensory Processes*, ed. R. Llinás and P.S. Churchland (Cambridge, MA: MIT Press, 1996): 29–60.

Chapter 14

1 Mark Greenberg and Martha Farah, "The Laterality of Dreaming," *Brain and Cognition* 5 (1986): 307–321.
2 Antti Revonsuo, "The Reinterpretation of Dreams: An Evolutionary Hypothesis of the Function of Dreaming," *Behavioral and Brain Sciences* 23 (2000): 793–1121.
3 Owen Flanagan, "Deconstructing Dreams: The Spandrels of Sleep," *Journal of Philosophy* 92 (1992): 5–27.

Chapter 15

1 A. Revonsuo, "Can Functional Brain Imaging Discover Consciousness?" *Journal of Consciousness Studies* 8, no.3 (2001): 3–23.
2 Timothy J. Andrews, Denis Schluppeck, Dave Homfray, Paul Matthews and Colin Blakemore, "Activity in the Fusiform Gyrus Predicts Conscious Perception of Rubin's Vase-Face Illusion," *NeuroImage* 17 (2002): 890–901.
3 N. Kanwisher, "Neural Events and Perceptual Awareness." *Cognition* 79 (2001): 89–113.

4 M. Raichle, "The Neural Correlates of Consciousness: An Analysis of Cognitive Skill Learning," *Philosophical Transactions of the Royal Society of London* B 353 (1998): 1889–1901.

5 Troels Kjaer, Markus Nowak and Hans Lou, Reflective Self-Awareness and Conscious States: PET Evidence for a Common Midline Parietal-Frontal Core," *NeuroImage* 17 (2002): 1080–86.

Chapter 16

1 Daniel Dennett, "Are We Explaining Consciousness Yet?" *Cognition* 79 (2001): 221–37.

2 Bernard Baars, "In The Theatre of Consciousness," *Journal of Consciousness Studies* 4 (1997): 292–309.

3 Bernard Baars, "How Does A Serial, Integrated and Very Limited Stream of Consciousness Emerge from a Nervous System That Is Mostly Unconscious, Distributed, Parallel and of Enormous Capacity?" in *Experimental and Theoretical Studies of Consciousness*, Ciba Foundation Symposium 174, ed. G.R. Bock and J. Marsh (Chichester, UK: Wiley, 1993): 282–304.

4 D.J. Watts and S.H. Strogatz, "Collective Dynamics of 'small-world' Networks, *Nature* 393 (1998): 440–42.

5 E. R. John, P. Easton and R. Isenhart, "Consciousness and Cognition May Be Mediated by Multiple Independent Coherent Ensembles," *Consciousness and Cognition* 6 (1997): 3–39.

6 Robert Doty, "Commentary on E.R. John, P. Easton and R. Isenhart, 'Consciousness and Cognition May Be Mediated by Multiple Independent Coherent Ensembles,'" *Consciousness and Cognition* 6 (1997): 40–41.

7 E. Bisiach and C. Luzzatti, "Unilateral Neglect of Representational Space," *Cortex* 14 (1978): 129–33.

8 Susan Blackmore, "There Is No Stream of Consciousness," *Journal of Consciousness Studies* 9 (2002): 17–28.

Epilogue

1 Quoted in Allan L. Smith and Charles Tart, "Cosmic Consciousness Experience and Psychedelic Experiences: A First Person Comparison," *Journal of Consciousness Studies* 5 (1998): 97–107.

2 Andrew Weil, "Pharmacology of Consciousness: A Narrative of Subjective Experience," in *Toward a Science of Consciousness*, ed. S. Hameroff, A. Kaszniak and A. Scott (Cambridge, MA: MIT Press, 1996): 677–89.

Index

Page numbers in *italics* refer to illustrations. Page numbers followed by the letter *n* refer to material discussed in footnotes.